RECONSTRUCTING RELATIVITY

An Inquiry into the Nature of Truth

Ted Cole

Book 1 – Special Relativity

This book is dedicated to the memory of my grandfather. When I asked him how far it was across the lake, he got out his surveying transit, a tape measure, and a book of trigonometry tables. Then he triangulated.

"Dreams of falling, dreams of flying,
 A man who never dreams goes slowly mad.
 The dawn of science, the age of reason…
 This is the voyage of the mind's eye"

       - Thomas Dolby
          from "Valley of the Mind's Eye"

Acknowledgements

Innumerable thanks are owed to Tony DeLyria for over ten years of support, intellectual challenge, and guidance on this project. Many thanks also to his daughter Joy DeLyria for her high-quality editorial advice regarding diagrams and drafts of the text. Thanks to Ricardo Robles for his supportive interest, and for his "turtles" remark. Thanks also to Margaret Bush for assistance regarding manipulation of graphic images in Microsoft Word. This project was triggered in 2008 by renewed interest in philosophy after connecting with Dr. Anthony McWatt via his website related to the works of Robert Pirsig. Thank you, Ant and Bob. Acknowledgements are due to all the writers and lyricists quoted in this book. I strongly recommend that readers seek out and experience each quoted source. Finally, thanks to my wife Joan for all her love, her patience, and her support for independent thinking.

Copyright 2019 Edward H. Cole Jr.

Contents

Book 1 – Special Relativity 1
INTRODUCTION 8
SPECIAL RELATIVITY - Part 1: Constructive Model 13
CHAPTER 1 - ABSENCE OF A CONSTRUCTIVE MODEL 13
CHAPTER 2 - HISTORICAL BACKGROUND 15
CHAPTER 3 - STEPS TOWARD A CONSTRUCTIVE MODEL 19
    FIRST STEP: PERPENDICULAR PLANE HELICAL MOTION 23
    SECOND STEP: PARALLEL PLANE RADIAL MOTION 27
    THIRD STEP: PARALLEL PLANE CIRCULAR MOTION 42
CHAPTER 4 - SYNCHRONICITY AND TIME SKEWING 47
CHAPTER 5 - THE MEANING OF NOW 57
CHAPTER 6 - SEPARATING SYNCHRONIZED CLOCKS 58

SPECIAL RELATIVITY - Part 2: Applications 68
CHAPTER 7 - THE PRICE OF ABSTRACTION 68
CHAPTER 8 - LADY ON A TRAIN 70
CHAPTER 9 - ON SPACETIME 72
CHAPTER 10 - TWIN PARADOX 74
CHAPTER 11 - SYMMETRY PARADOXES 83
    DILATION AND CONTRACTION SYMMETRIES – AN EXAMPLE 84
    PRELIMINARY STEP: ESTABLISHING TIME SKEWING IN CLOCKS 85
    TAKING ON THE PARADOXES 87
    SYMMETRY OF MEASURED RELATIVE SPEED 88
    SYMMETRY OF CLOCK RATES 89
    SYMETRY OF LENGTH CONTRACTIONS 90
CHAPTER 12 - LORENTZ TRANSFORMATION EQUATIONS 92
AFTERTHOUGHT - PULSARS: AN ABSOLUTE TIME REFERENCE? 100
THOUGHTS ON THE SOCIAL IMPLICATIONS OF RELATIVITY 101

PREFACE

Anthropomorphism is easy to dismiss as sloppy thinking, but it contains a larger concept worth examining. Attribution of specifically human thought processes to other lifeforms or inanimate objects is erroneous. But our sense of likes, dislikes, values, and preferences are just human manifestations of a larger principle that applies to all things: non-living and living things, plants, animals and people, social and cultural and political things, and intellectual things. To say that "rocks like to roll downhill" or "Oxygen likes to bond with Hydrogen" is not to treat these non-living things as if they had a brain and human emotions. Rocks and atoms are not "like us"; we are like them.

Like many people, I have searched for ways to make sense on all levels of the Universe in which I live. What is the best explanation? What models best fit reality as I experience it? One of the best all-inclusive models I've found comes from the ideas of Robert Pirsig, author of "Zen and the Art of Motorcycle Maintenance" and "Lila." My revisiting of Pirsig's books in 2008 inspired the journey that resulted in this book. Pirsig suggested that Quality, value, the sense of "what's better" is the central player in the evolution of everything. In his model, quality has two modes: static quality, including all that has existed and endured - all of which exists only in the past, and an undefinable dynamic quality which exists only in the present and is on the cutting edge of the future. Things exist and endure because they have been better than the other options, the other structures, the other things that were not sufficiently robust, that were not as good.

Everywhere I look and on every level of existence I see this interaction between the static good and the dynamic good. It can be simplified to this: 1) Try a new thing. 2) Is it better? If so, keep it and make it a part of the stable pattern. If it is not better, stay with the existing best stable pattern. Quantum Mechanics: spontaneous creation of particle and anti-particle from the vacuum… No, not better, recombine and return to the average nothingness. Chemistry: opportunity to break chemical bonds and form new ones? Yes, the new one is better, stay with that. Or no, go back. The origins of Biology: what chemical bonds are stable? What elements exhibit properties which both allow them to try new things, and yet allow them to form stable structures? These are the elements that have developed into living structures. What structures can endure? What new survival advantages can be found? Those that were not as good did not endure, and surely there were many failed attempts.

Biologically, with certainty, the evolutionary process exhibits the character of "locking in" the static good, dropping the inefficient or unnecessary, and striving to find a new advantage, whether sensorial, structural, or intellectual. Eagles have very good eyes because, once there were eagles with "not so good" eyesight, but they did not catch as many rabbits as the eagles with better eyes. One could say eagles "like" to have good eyesight. It's their preference, it works for them. For us humans, we see the same static/dynamic quality processes in our lives personally, socially, and intellectually. Biologically we evolved memories and the capacity to create language. We like to have memory and language, they have worked for us and we have kept these things.

With human memory and language, and learning, we have a complex relationship to quality that involves time. What is good short term and what is good long term? Is a long-term stable family better than short-term relationships and no family? How can we best teach our children so that they may live well? How can we best structure our tribes, governments and societies so that the tribe or society or government is best and so that it also survives to be good for a long time?

In politics, generally, conservatives value maintaining the stable good, the static quality. Liberals and progressives value trying new things which promise to be better than the status quo. Rapid dramatic change, such as a social or political revolution, threatens to destroy the existing structure, which if it fails, can lead to something worse. But prohibiting change means prohibiting freedom, which precludes the chances for things to get better. These days, it seems to be the rare centrists who are pragmatic. They are willing to try something new, but if it doesn't work out well, admit the mistaken direction and try something else, or return to the prior mode of operation. As a society, we fluctuate between these two poles of thought and practice.

At a higher level, the same concepts apply to intellectual activity, to thought itself. Science is usually conservative, seeking incremental and tentative advances in knowledge. But sometimes, there are revolutions: The Copernican revolution, Newtonian mechanics, relativity, and quantum mechanics. These revolutions in thinking all caused turmoil and chaos which gradually affected the societies and cultures from which they arose. We are living now in the wake of the relativity and quantum mechanics revolutions, and we're still digesting the implications. A significant fraction of my life has now been devoted to digesting relativity: what it is, how it works, and what it means.

Fundamentally, I am a temporal naturalist. I believe that time is a real thing. The Universe and the things in it are real, not imagined. Mathematics is a useful tool for describing reality; mathematics is not reality itself. The universe proceeds from moment to moment. The future has not happened yet. The past has happened. Though attempts to record and reconstruct what exactly happened will always be imperfect, I believe that the past only happened one way. The uniqueness of the past therefore constitutes an absolute, though not perfectly knowable, truth.

As will be seen, this philosophy is at odds with spacetime relativity, in which past, present and future, and position and velocity in space are all tied together in a mathematical construct called space-time. In traditional relativity theory, time is not "real" in the intuitive sense, it is simply a measurable entity. In traditional relativity, mathematics is presented as the best representation of reality, and truth itself is considered relative.

- Houston, 2019

INTRODUCTION

Thinking is a momentary dismissal of irrelevancies.
                                  - Buckminster Fuller

There has been a philosophical debate which dates back to the Greeks regarding whether truth is relative, or truth is absolute. From the time of Newton to the early 20th Century, the dominant western philosophy was that truth was absolute. Space and time were absolute. Physical phenomena were considered determinate; if there was a lack of certainty, the only deficiency imagined was an inability to gather a sufficient amount precise data. This view of reality dominated western thinking until Einstein. Relativity established truth as a relative entity. Space and time are no longer considered universal absolutes. Different observers can perceive events differently and both observers are considered to be correct. Even the sequence of events can be perceived differently, and both observers are still considered correct. Their perception of the facts is relative - it depends on their state of motion.

The very word relativity defines this philosophy. The different observations of reality which multiple observers can all correctly hold is defined by their motion, but not by their motion relative to some absolute universal frame. Einstein says that their different perceptions of reality are defined entirely by their motions relative to each other. As Einstein and modern Physics sees it, there is no absolute reference frame. There is, therefore, no absolute truth. Different observations of reality are all held to be equally valid.

My recent process of thinking about relativity was not initially directed toward this debate about the nature of truth. I did not set out to change anything. I just wanted to understand relativity, because I had tried to understand it as a young man, and I had failed. As an engineer, I had no need to understand it. For all practical purposes of mechanical engineering, relativity is unimportant. Newton's mechanics works just fine for just about everything mechanical engineers deal with. I was curious about relativity because it had been mentioned briefly in high school physics, and it seemed fascinating and mysterious. So, I took a course in relativity as an elective option in college. I passed the course, but I never felt that I understood how it really worked. Though I rarely ever talked to professors outside classes, I sought out this college physics teacher and tried to express my confusion. He smiled and chuckled kindly, but he really had no answers for my questions. The sense I got from him was that either he didn't really understand it himself, or that it was way too deep for a dabbler like me to comprehend.

When I returned to thinking about relativity in 2008, I had been in a reflective and philosophical mood; I was thinking about everything I had ever learned and experienced. Everything seemed to fit together and make sense. I experienced an enormously harmonious feeling about so many things – until my mind stumbled across relativity, and a dark cloud came over me. Why did I have this vague discomfort every time it crossed my mind? Why did it not seem to fit with the beautiful harmony that I felt about everything else? I decided to try again to understand relativity.

I got some books, and started learning about it again, but soon I ran into the same old confusions. Only this time, I was not in a course. Now, there was no time limit, and no test except the test of my own satisfaction. I wanted to bring relativity into a harmonious realm of understanding with everything else. But I could not find a book which explained it to my satisfaction, so I was forced into the unimaginably painful position of having to think about it myself. I had to think really hard. For a time, I was obsessed, and for one brief period of several weeks, I believe I was technically insane. I could not stop thinking about it, even when I wanted to. I took long walks, often to a fish hatchery near my home. I watched the circular ripples in the ponds. One day after a rain, I watched a series of raindrops fall from the roof of the covered bridge into the slowly moving river below. Were the rings these drips made in the river circular? Yes, maybe… circular, but moving – circular, but not concentric? I watched bugs circling a light in the evening. How would a light breeze affect their motion? I imagined moving swarms of bugs pulsing rhythmically in the air. What would this look like, as a swarm moved past me? What sort of paths would the individual bugs take?

Months followed with uneven progress, but I began to put the pieces together. What I had been looking for, and what I could not find in the books I was studying, was a PHYSICAL explanation of the phenomenon of relativity. The reason I could not find it, I discovered, was that Einstein did not give a physical explanation for it in the first place. And no one had seemed to offer one since. What Einstein provided was a mathematical description of how several observers' measurements of the same events would be different based on their motion relative to each other. But Einstein's proof began with a postulate which was based on experimental evidence of a phenomenon which was not understood and which he never physically explained.

Johnny Carson had a saying which comes to mind in describing his comedy. Paraphrasing, he said that "if they buy the premise, they will buy the bit." In retrospect, this fits Physics' description of relativity. When I followed Einstein's derivation of the equations that define Special Relativity, I could find no flaw. I accepted each step from the beginning to the conclusion. What I finally came to realize was that the premise, Einstein's initial postulates, contained the mystery. I was years into my recent study of relativity before I discovered that Einstein had despaired that he could not come up with a 'constructive model' which explained the phenomena which underlie relativity. Instead, he settled for basing his theory on a 'formal principle.' This formal principle was a scientifically proven fact which he knew, or strongly believed, could not and would not be disproven. There was already plenty of evidence which bore it out. This fact was that the speed of light in a vacuum will be measured to be the same by any and all observers, regardless of the observer's own speed through the vacuum. Einstein called this the Principle of Relativity. Once you buy that premise, all of relativity follows. If you buy the premise, you buy the bit.

My purpose here is not to dispute the premise. My purpose is to explain it, physically. I want to begin to build that constructive model which Einstein did not build. To the extent that a rational constructive model can be developed, harmony may return, and the phenomena of relativity may make sense. One unforeseen effect of building such a constructive model was the impact it would have on the debate originally discussed, the debate about the nature of truth. To build a constructive model, it appears that we need to return to some notion of absolutes.

This notion of absolutes, however, is not a return to the Newtonian sense of absolute space and time. Newtonian thinking supposed that space and time were absolute in that all observers would be in agreement, that all observers' sense of time and distance would concur. Everyone's clocks would always run at the same rate, without question, to the limit of their precision. That is not what I am proposing. The absolute space and time which I am thinking about is an absolute upon which all observers will not be able to agree. It is a notion of an absolute which is not knowingly determinable by any observer. It is a notion of absolutes that are beyond the philosophical program of subjects and objects, beyond scientists and their measurements.

I love science, and I appreciate that it has accomplished so much. Science has been enormously successful because of its Primary Value: the thing it holds dear above all else is Repeatable Experimental Measurement. Without that, reasoning about how things work is largely just speculation. Aristotle *reasoned* that heavier objects must fall to the ground more quickly than light objects, but he did not do the experiments to validate his reasoning. His wrong ideas lasted for centuries. But at the dawn of Science, Galileo did the experiments, and proved Aristotle was wrong. Bravo. The scientific method of proposing a hypothesis, then testing the hypothesis repeatedly until a near-certainty could be established led to a watershed of advancement in human understanding and technological progress.

But despite all the indisputable gains of science, was something also lost? What I think might have been lost was the awareness that science was still a philosophical pursuit. After all, the certificate awarded to all those who claim real expertise in science is the PhD; they are all Doctors of Philosophy. Science is a philosophical pursuit which values measurement in its attempt to understand reality. Measurements involve an experimental Subject (the experimenter), observing an experimental Object. This works extremely well as long as the object being observed is sufficiently different from the subject so that no confusion or conflict arises about which is which, and it can be reasonably said that one has little effect on the other. This is clearly the case with scientific experiments about dropping balls, steam under pressure, beams that are deflected under loads and thousands of other phenomena. For the entire period from the Renaissance through the Industrial Revolution, the assumption that the scientific Subject can objectively observe the studied Objects worked extremely well.

But this assumption does not work so well when the distinction between subject and object is not so clearly definable. For example, the subject-object distinction is not as clear when scientists, who are people with cultural backgrounds and human behaviors of their own, set out to take measurements on the social behavior of other people. The objectivity of the scientists can be subject to ongoing debate in a way that simply doesn't occur when properties of a gas or the deflections of a structure are measured. This does not mean that the Social Sciences have no value – they do. But any good social researcher or Psychologist knows that they need to examine themselves as well as the people they are observing. Social researchers cannot entirely separate their own humanity from the humans they study.

The subject-object model also is not perfect when people who are composed of molecules and atoms and sub-particles set out to do experiments on atoms and sub-particles with measuring instruments that are composed of atoms and sub-particles. In such experiments, some subtle issues may arise which need to be carefully considered before drawing definite conclusions. If on the most fundamental level we are made of the same stuff as light, whatever happens to light also happens to us. The subject and the object are, in an important sense, One not Two. This should not prevent us from thinking about the universe we see, but we should be very careful about our premises and our scientific process. We should consider exactly what a measurement is and include awareness of this analysis in our philosophical valuation of measurements above all else.

What I am proposing is the philosophical notion that there is one and only one true reality, an absolute truth, which we might call 'the evolving state of the universe' which cannot be objectively measured because we cannot remove ourselves from the universe to make an objective measurement. This absolute truth, I suggest, is a reality which no intelligent physical observer can claim to definitely, absolutely know. To propose this notion is a philosophical pursuit which scientists are almost certainly bound to reject because it cannot be tested with a scientific subject-object experiment.

But this is not a scientific proposal, it is a philosophical one. The test of such a philosophical proposition is not whether it is physically provable by measurement, but whether it is philosophically more harmonious than the current philosophical model, represented by relativity as it is currently viewed. Of course, any new alternative philosophical model, to be credible, must remain consistent with experience. The new model must still fit the existing data. That is, in this case, it must be consistent with more than a century's worth of experimental evidence. But if an imaginable constructive model can be demonstrated which is consistent with the facts, refers to a notion of absolute truth, and also presents more philosophical harmony than the current notion of relative truth, I suggest that the debate over then nature of truth should not be considered closed.

Einstein said that there is no preferred reference frame, which simply implies that no single observer can claim to know absolute truth. But because science values measurement above all else, science reached a critical juncture with relativity. Because multiple observers traveling at different speeds relative to each other will obtain different measurements of the same objects and events, including measurements of an object's mass and size, science decided that each observer has his or her own personal truth, his or her own personal reality, and all of these different truths are all equally valid.

The reason Einstein rejected the concept of the uncertain absolute is that experimentally, the 'absolute frame' could not be located. At least, it could not be located from within a windowless laboratory. But beyond that, Einstein liked the notion that the Laws of Physics should be identical for all observers, independent of one's choice of reference frame. He saw an elegant symmetry in that notion. Speculation about an 'absolute' spatial reference frame was moot, he said, for all ***practical*** purposes.

And there's the key: practical purposes. Einstein's practical purposes involved measurements and predictions. I will argue that there is at least one practical purpose for which the concept of the existence of an absolute frame of reference is useful – the practical purpose of understanding what is physically happening, for the practical purpose of making relativity make sense. Although an absolute frame of reference has not yet been experimentally located*, it can still be conceptually imagined. While agreeing that such a 'non-moving' frame may not be 'preferred' over any other, it should also not be banned from consideration, if it provides some value in helping us imagine the physical phenomenon in our mind's eye.

I've found that for "practical purposes" such as relating measurements in one moving reference frame to measurements in a different moving reference frame, imagining or establishing a 'non-moving frame' is purely a pain in the butt. Why in the world would anyone want to refer data in a real moving reference frame A to an (only imagined) stationary reference frame B, and then from frame B to a different real moving reference frame C, when computations can take you from frame A to frame C directly, in one easy step? Indeed, it's impractical, and a waste of time, and completely unnecessary.

But the imagined stationary reference frame B happens to be the one place you can go in your mind's eye if you want to 'see' what is happening physically. And that's what I want to do. I want to slow down, back up, and look at the big picture. I want to burrow down into relativity and see what's really going on, to make sense of it. I want to de-mystify it and make it understandable by every person who cares to think about it, not just by the experts.

---

* Physics professionals will likely tell you an absolute frame cannot be located – at least, not from within a windowless laboratory. Later in the book, I explore a conceptual 'thought experiment' (with windows!) designed to locate a reference frame which is not moving through space locally.

SPECIAL RELATIVITY - Part 1: Constructive Model

CHAPTER 1 – ABSENCE OF A CONSTRUCTIVE MODEL

Ring the bells that still can ring.
Forget your perfect offering.
There is a crack, a crack in everything.
That's how the light gets in.
        - Leonard Cohen, Anthem

There is a way in which relativity as taught by Physics and as explained by most authors is not just complex: it's crazy. And that's the part I want to focus on – the crazy part. It shouldn't make us feel crazy. It should make sense. We shouldn't have to live with something that feels wrong or gives us a queasy sense of disjointed reality. We also shouldn't have to feel humiliated or ashamed when it is implied that if we don't understand it as they explain it to us, the reason for this is that we lack the mental capacity to do so. It can't be THAT complicated. No good theory should be conceptually incomprehensible.

Relativity, if we understand it right, should give us a good, harmonious feeling that we understand reality better than we did before we learned about it. When we add relativity to our mental toolkit, our view of the universe should be corrected and adjusted, not shattered. The purpose of this book is to present an alternate explanation of relativity that is comprehensible and fits comfortably within the realm of common sense. This alternate explanation is not meant to replace the standard approach, but to augment it. There will not be any new results here; Einstein's relativity has been verified in countless laboratory experiments.

But what should we do if there are two ways of explaining something, both of which are consistent with experiments? What if one explanation is mathematically expedient but philosophically jarring, and the other explanation is philosophically harmonious but mathematically inefficient? Perhaps we should allow both interpretations: using the harmonious model for physical understanding and using the mathematically efficient approach for practical problem solving.

Mathematicians will no doubt argue that mathematical elegance *does* imply philosophical superiority, and there is a legitimate argument to be made in that regard. However, when mathematical abstraction tends to conceal the physical explanation, an argument can be made that Physics has deviated from one of its main purposes, that of physical explanation.

I think that what has happened with relativity is that the mathematically expedient interpretation has been accepted by Science and Physics in particular as the *only* valid interpretation. In other words, the "mathematically practical" has been accepted as the

physical reality. This is probably due to the historical development of relativity, and the general de-valuing of Philosophy by Science in general because many philosophical arguments involve un-measurable constructs. This is a blind spot which Science seems to have, perhaps as an artifact of the history of Science as a whole. Much like the blind spot in our visual field, it's small and off to the side, easily patched over, and it never seems to interfere with the primary vision of Science.

Constructive Model vs. Formal Principle

"When we say that we have succeeded in understanding a group of natural processes, we invariably mean that a constructive theory has been found which covers the processes in question. . ."
- Einstein, 1919

"By and by I despaired of the possibility of discovering the true laws by means of constructive efforts based on known facts. The longer and the more despairingly I tried, the more I came to the conviction that only the discovery of a universal formal principle could lead us to assured results."
- Einstein (autobiographical notes)

When Einstein spoke of his inability to "discover the true laws by means of constructive efforts," what did he mean? I think he meant that he did not have a satisfactory physical explanation for the phenomenon of relativity. He implied that on some level, he did not understand how it works. He could not "see it" in his mind's eye. But knowledge of the atom and smaller particles was rather lacking at that time compared to today. Had Einstein approached the problem with a fresh mind a few decades later, I think the outcome may have been different. If a constructive model had been imagined from the start, things might have also gone a little differently regarding how the subject is taught and explained. Subsequent attempts to explain relativity to the general public without a constructive model have been disappointments at best and complete failures at worst.

In this book, we will go back to the very beginning. In fact, we will go back to the time before Einstein's first paper on relativity, to develop a constructive model which will give physical meaning to the formal principle which he used as his starting point. Our constructive model will explain *why* Einstein's formal principle was correct. This should fill in the missing piece of the puzzle.

We humans are storytellers, we dream, we imagine, we see pictures with our mind's eye. When we try to understand something that we do not yet understand, we think metaphorically. We form mental pictures that make connections between the familiar and the strange. This metaphorical thinking comes naturally. It is the basis of our language. It is literally the way our minds work. When we look up an unfamiliar word in a dictionary, we usually find it defined by words with which we are more familiar, but if some of these words are also unfamiliar, we look up those words too, until we finally come to words which we already understand.

This book will use such metaphorical thinking so that we can imagine how relativity works. This activity is not fantasizing, however. We will repeatedly come back to rationally evaluate whether the metaphor is working. Does it still fit with the experimental evidence? Does it get the same results as the abstract relativity mathematics? As long as the results match reality, and the mental image is satisfyingly simple, we can claim to be making progress in understanding.

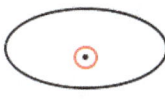

Without a constructive model, relativity remains esoteric, abstract, and something best left to the mathematical experts. Without a constructive model, relativity is literally unimaginable. In this book, let's begin again to imagine.

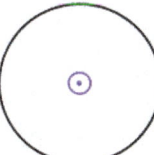

CHAPTER 2 - HISTORICAL BACKGROUND

"The modern concept of the vacuum of space, confirmed every day by experiment, is a relativistic ether. But we do not call it this because it is taboo".

- Robert B. Laughlin, Nobel Laureate in Physics,
endowed chair in physics, Stanford University, 2005

Despite his lack of a constructive model, Einstein was enormously successful in propelling relativity theory forward in a mathematical context that was correct, and very useful. How can this possibly be? As he stated, Einstein fell back to reliance on a 'formal principle,' which means that he took as a starting point a fairly well-established fact which he could not fully explain. Supposing this fact was true, he then asked, what would be the logical outcome in terms of observed measurements?

He called his principle the Principle of Relativity, which stated that the Laws of Physics should be the same in all reference frames - that there is no 'preferred' reference frame. Maxwell's breakthrough electrodynamics theory (ca. 1862) implied that light, like other electromagnetic waves, moved at a constant speed c in a vacuum. But in what reference frame was such a speed to be measured? If two moving reference frames measured the speed of the same light wave, would the speed measured be different for the two measurers? Maxwell's equations did not implicitly refer to an absolute reference frame, which left the implication, if Maxwell was right, that the same speed c would be measured by observers in any moving reference frame. Experiments at the turn of the century (1900) were all confirming Maxwell's equations.

The notion, specifically, that the speed of light in a vacuum would be measured to be the same in any reference frame was the essence of Einstein's 'Formal Principle' of Relativity. When you think about it, this is a truly amazing fact. It means that two different observers who are moving at a very high rate of speed relative to each other would both measure the same flash of light as having the same speed relative to themselves. This seems to defy common logic. This is different than all other experiences we have with relative velocities and wave phenomena.

For example, if someone on the ground measures an acoustic wave as moving at 767 miles an hour to the right, then someone in an airplane moving at 200 miles an hour toward the right will measure the speed of the acoustic wave as 567 miles an hour relative to themselves. The velocities are considered additive. A baseball thrown forward at 70 miles an hour on a train which is moving 30 miles an hour will be measured as moving 100 miles an hour by an observer on the ground. This is the familiar Galilean/Newtonian rule of additive relative velocities.

Not so, with light. Someone on the ground will measure a flash of light (in a vacuum) to be moving at 299,792,458 meters per second toward the right, but someone in a super-high-speed rocket moving at 200,000,000 meters per second toward the right will NOT measure the speed of the same flash of light as 99,792,458 meters per second relative to themselves, they will in fact also measure the speed of the same flash of light as 299,792,458 meters per second relative to themselves. That is an amazing fact which has been born out in repeated experiments.

I think that the question of why the speed of light is measured to be the same in all moving frames is what caused Einstein despair. He could not come up with a 'constructive model' that explained it physically. But he knew experiments confirmed it. So, he set forth the 'formal principle' of relativity, which is that there is no preferred reference frame for measurements, and that the laws of Physics should be the same regardless of the motion of one's reference frame. And primary among the laws of physics is that the measured speed of light in a vacuum is a fundamental constant.

What Einstein accomplished was a resolution of the apparent contradiction which this fact caused. To do this, he had to throw out Newton's concepts of "Universal Time" and "Universal Distance." Time and distance became for Einstein not fixed things, but variable quantities which could stretch and shrink in an odd way. This led to an entirely new dynamics which replaced Newtonian dynamics. For example, in the example given above of the baseball thrown from a train, Einstein's new dynamics says that the velocities are not precisely additive, as Newton had predicted. With Einstein's new dynamics, the baseball would be measured as moving slightly less than 100 miles an hour if measured from the ground.

But the deviation from simple addition is extremely small for measured speeds which are very much slower than the speed of light. In other words, Einstein's dynamics coincides with Newton's dynamics for all practical purposes regarding objects moving at normal everyday speeds, and I mean this to include baseballs, bullets, supersonic jets, and NASA's spaceships traveling to the moon. Einstein's dynamics significantly deviates from Newton's dynamics only when the speeds involved are a significant fraction of the speed of light.

At very high relative speeds such as the two observers mentioned above on the ground and in the extremely fast rocket, these "relativistic effects" would become significant. These observers would both measure the speed of light relative to themselves to be the same, but when measuring things other than the speed of light, their measurements would be different. They would get different answers regarding the speed, shape, and mass of moving objects, the time which passed between two observed events, and even in some cases the sequence in which events occurred. And given that the primary mystery about the speed of light was a true fact, Einstein could correctly predict the discrepancies in all the other measurements they would make.

This gave physicists a whole new toolkit for making and interpreting their measurements, especially in the realm of high-speed particles observed in physics labs. Yet the primary mystery remained physically unexplained. There was no constructive model, only the formal principle. The historical unfolding of relativity and 20th century physics is crucial to an understanding of why no constructive model was ever developed. I believe a constructive model is available, but no practicing physicist ever dared to put one forward, or if they did, they were shot down, because of this history.

Before Einstein's paper on Special Relativity was published in 1905, the Dutch physicist Lorentz had tried to explain the experimental results that were showing that the speed of light was the same as measured in all reference frames. The equations that Lorentz produced in this effort are identical to Einstein's equations, and in fact, Einstein's initial relativity equations are still called the Lorentz transformations, because Lorentz came up with them first. But when Lorentz put a physical explanation forth, his physical explanations for the phenomenon were roundly rejected.

Lorentz just didn't have quite enough information to assemble a plausible explanation at the time, and so his constructive models were considered 'ad hoc' – which is not a good thing, to scientists. Lorentz's explanations seemed contrived to make the experimental results 'come out right' without a valid reason why. Lorentz described 'length contractions' necessary for the experiments, notably in the Michelson-Morley experiments, to produce the results they did. His explanation was that the 'ether' that filled space (the presumed light-wave bearing medium) somehow compressed solid objects in their direction of their motion through it. But he had no specific reason for why this happened and no specific explanation for the amount of compression involved. Worse than that, as part of his 'ad hoc' package, Lorentz produced an equation about time that he had little explanation for, except that it was also required to explain the expected results if the Michelson Morley experiment were to be set up a little differently. Clocks, he said, would have to run at specific different rates, and be offset in time from each other based on their physical locations in a specific way.

Rather than trying to explain the physical phenomenon (or perhaps after trying and not being able to visualize one) Einstein in essence said: We have been beating our heads against a wall trying to understand this, but let's stop doing that, and simply accept the experimental result as the truth. Then we can go forward. And by taking that approach, he was able to move forward, mathematically. The speed of light is measured to be the

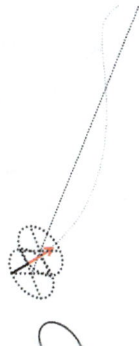

same regardless of the linear speed of one's frame of reference when measuring it. Sure, that's what nature is telling us, he said. We simply have to recognize it and recognize the implications and ramifications of this on our previous assumptions of how mechanics operates. And this put a very significant ding in the previously assumed perfection of Newton's Laws. That's huge. In doing this, Einstein made the mechanics of solids and electrodynamics compatible.

Einstein therefore was considered to have the conceptual breakthrough which Physics needed. Einstein had the Principle of Relativity. With that, many problems were resolved. All the experimental data fit. Everything seemed to make sense; everything, that is, except the original mystery. Why was the Principle of Relativity true? How did that work, physically? No one seemed to care much about that. And to Einstein, the Principle of Relativity did have a certain elegance of its own – this idea that all observers should experience the same thing, that the Laws of Physics do not have a preferred reference frame. That formal principle seemed to be satisfying enough.

Lorentz was trying to explain the phenomena in relation to the 'ether' which was imagined as the medium which bore the light waves. But to Einstein, and other scientists of the day, this speculation about this unmeasurable, undetectable 'ether' was of little value. It was literally in the realm of Philosophy, not Science. With Einstein's approach, which relied only on experimental evidence, the ether was moot, the existence of a medium in which the waves travelled was considered unnecessary, and the concept of a wave-bearing medium disappeared. All that mattered for the purpose of calculations was the relative speeds of the frames of reference under consideration. Certainly, with the wave-bearing medium removed from the picture, further attempts at a physical explanation were impossible.

Shortly after Einstein's breakthrough, his mentor Minkowski took relativity to a higher level of mathematical abstraction. Minkowski realized that the Lorentz transformations could be interpreted with a matrix mathematics that greatly simplified calculations. This mathematical formulation was called space-time. While very efficient, Minkowski's space-time mathematics further removed Physics from consideration of physical explanations and moved it more towards abstract mathematics. Physics never looked back. Special Relativity is the name given to the very limited consideration of constant speed, straight line motion. When Einstein moved on to develop General Relativity, which includes gravity and acceleration, he did so entirely in the mathematical space-time framework, and again, no real attempt was made toward giving any physical explanations. In the early 20th century, up-and-coming young Physicists didn't want to look back, they wanted to keep up, and if possible, get to the front. The winners in this race were all very good mathematicians.

But I think it is worth looking back. I think that other developments in Physics since the time of Lorentz and Einstein's early days have provided enough clues to start to build a valid constructive theory for relativity. And I am sufficiently removed from professional Physics that I have little at risk in doing so. Many authors have expressed displeasure with the way that Special Relativity is explained in textbooks and in dummied down versions meant for public consumption.

I have only encountered a few authors who I feel have been on the right track regarding a constructive physical explanation of Special Relativity; I found these only after developing my own constructive model. While finding people of similar mind was encouraging, I still found the texts of these authors lacking in different ways, and I wondered if these texts would have been adequate or even helpful to me if I had not already had a constructive model of my own. I offer what I hope is a better explanation. I want to write the explanation that I wished had been available to me long ago.

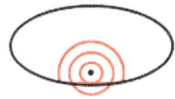

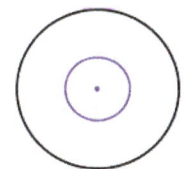

CHAPTER 3 - STEPS TOWARD A CONSTRUCTIVE MODEL

"…and in this struggle
it's important that
you know that you are fine.
The world is handout on the corners
and the burden of proof is on you."

- Margaret A. Roche, <u>Burden of Proof</u>, from "Seductive Reasoning"

Rather than start, as Einstein did, with a formal principle which says that light is measured to be the same by observers in all moving frames, let's start with a different postulate which is consistent with experiments, and try to build a working physical model that explains the physical phenomena of relativity. We have information which Lorentz and Einstein did not have when they tried to construct a physical explanation early in the 20th century.

What I'm about to propose as my fundamental postulate is not a new idea. It's not wild speculation, and it is basically aligned with modern theoretical and experimental physics. What may be unusual is the selection of this proposition as a starting point for understanding relativity. One of the major advances in 20th Century physics has been the development of the concept of wave-particle duality. Light, commonly thought of as a wave phenomenon, also sometimes behaves more like a particle. Matter, on the other hand, usually thought of in terms of particles, has also been found to exhibit wave properties. Not just some matter, all matter.

All indications are that, despite perceptions on our human scale, when you get down to a very small scale, there are no solid little nuggets of stuff. The things we have been calling 'particles of matter' are in fact not solid in any ordinary sense. They only seem solid because of their chemically bonded structures, and the mutually repulsive electric fields of the electrons that surround them. As we've looked closer and closer at particles of mass, and their component sub-particles, a new picture has emerged which is telling us that the variety of particles we observe is nothing more than a variety of patterns of energy.

While this failure to find any 'nuggets of stuff' as we go smaller and smaller is a strong trend, one could argue that the end of the trail has not been reached. At one point, we thought atoms were nuggets. Then we explored their structure and found protons, neutrons, and electrons, and they were at first thought to be nuggets. Currently, 98% of the externally observed mass of protons and neutrons are said to be owing to the energy within these particles, with only 2% of the mass attributable to the quarks traveling within them. Now, are we to believe that quarks are the new nuggets?

Physics continues to generate new theories regarding the ultimate structure of matter on the smallest scale. String theory suggests that the matter is an extremely tiny vibration of energy, with no nuggets in sight. But so far, no experiments have been proposed that could verify or disprove string theory. The inner structure and composition of particles of mass is not yet resolved with a comprehensive theory that has been confirmed by experiments. But the history suggests that there may indeed be no 'solid nuggets of stuff' whatsoever.

I offered this idea to a friend and co-worker, and he replied, "Oh, so you think that it's turtles all the way down?" Yes, that is what I'm suggesting. Turtles all the way down. No little nuggets of whizzing stuff. To say this is true, that all mass is energy which is trapped in a pattern, IS a postulation on my part. This is a postulate that could not have been reasonably imagined at the dawn of the twentieth century - but is quite imaginable today. When I propose the idea that 'there are no solid nuggets of stuff', I'm suggesting that the quarks too are composed of energy. If this has not yet been proved convincingly, I propose that the burden of proof should now be shifted to anyone who claims that the quarks, or any other particles, ARE indeed 'solid little nuggets of stuff.'

Energy, or whatever other word we want to use for it, seems to be all that there is, physically. Energy is distributed throughout space, and its distribution changes with time. Some energy runs freely through space, and this is the energy we have always been calling energy. The stuff we have been calling mass particles, however, seems to be energy that is trapped in very small coherent and persistent patterns. There are many different stable energy patterns, meaning there are different ways that energy is trapped. Each different pattern which contains energy is a different type of particle. The pattern defines the particle and the properties the particle exhibits externally. The fact that energy can be converted from the trapped particle 'matter' form into free-running incoherent energy is represented, in fact, by Einstein's formula $E=mc^2$. The amount of trapped energy is proportional to the externally perceived property known as the mass of the particle. Since Einstein didn't specify that only *some* types of mass can be converted to energy, it makes sense that *all* types of mass are composed of energy, and the 'conversion' of mass to energy is in fact a release of the energy trapped within the mass.

Now we are close to the center of the constructive model. I won't try to explain how the energy in a particle of 'matter' is trapped, nor will I try to describe the various patterns in which the energy is trapped. The beautiful part of building a constructive model of relativity is that we don't need to know the exact forms of these patterns. They may be circles, figure 8 loops, three-D loops, or something more like the standing wave of a vibrating string. A vibrating, or cyclic pattern that maintains its pattern is a repeating pattern. To not repeat the pattern is to lose the pattern. But again, we don't need to describe the patterns here. We just need to propose one idea about these patterns, and it's an idea that I think will be quite plausible.

That idea is that whatever the 'particular' patterns are that trap energy as 'matter', the energy moving within that pattern always moves at the speed of light through its local space. In other words, whether energy is in a free-running form, or in a trapped form, energy *always* moves through space at the speed of light. After all, at what other speed should we expect it to move? When we get down to the very fundamental level, we begin to conclude that "All Is One," that patterns, and relationships of patterns, are everything. We see that reality is composed of space, energy, and time, and all the variety of 'particles' and structures that we see are simply relationships of various stable and unstable patterns of energy transforming through time.

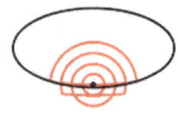

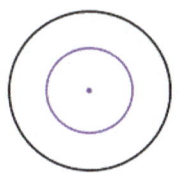

When we get to this very fundamental level, it makes sense that there is only one speed, for if there were more than one speed, we would suspect that there was some other fundamental component of reality which determines what the speed is, and we would suspect that we were missing some more fundamental component of our conceptual model of reality. No other 'fundamental speed' is known - just the speed of light.

This is the starting point then: Fundamentally, energy, space, and time are everything, and energy always travels at a uniform speed through space; this speed is a fundamental property of our universe. It defines the fundamental relationship between energy and space. With this, we will be able to envision how relativity works physically, not just mathematically. We will be able to develop a constructive model of relativity which will always work, despite our imperfect knowledge of all the patterns and forms which energy can take or what mechanisms trap the energy into the stable patterns we call particles. We also don't need to know how various particles interact with each other in order to discuss relativity.

It is important to note that in imagining this physical model, I have so far not mentioned any reference frame, nor have I assumed any 'observer' who is measuring anything. Einstein's starting point was quite different. Einstein was a scientist, and he needed to have his theory accepted by the body of scientists. Science values repeatable experimental measurement above all else. So, Einstein began with the formal principle that all measurements of the speed of light in a vacuum in the context of a scientific reference frame were numerically identical, and he reasoned forward mathematically to show that this assumption led to a new physics which correctly predicted measurements of objects at very high speeds – his new dynamics.

This new description of dynamics led to some very counter-intuitive results, including:

1) Observers in different reference frames with a mutual relative speed will **both** measure that time is progressing more slowly in the "other" frames than in "their own" frames.

2) Observers in different reference frames with a mutual relative speed will **both** measure that the lengths of objects in the "other" frames are shorter in the direction of motion than the lengths are measured to be by those in "their own" frames.

3) A traveler going on a very long trip through space at very high speed and then returning to Earth would experience fewer days than one who stayed on Earth, and therefore the traveler would age less than one who would stay at home on Earth. (The Twin Paradox)

For me, and I think for many people trying to learn about relativity, this Twin Paradox is the haymaker – the knock-out punch. It's one thing to accept that measurements of things such as clock rates and lengths of objects might be *distorted* due to the speeds of the objects involved. But the Twin Paradox is saying that there is a *permanent physical effect* which remains long after the measurements are made. And there is nothing in the traditional development of relativity that physically explains the cause of this permanent effect – it appears instead as simply a mathematical result of the derived equations.

While Einstein's development of relativity theory from his formal principle comes to these conclusions which have been validated by experiments, these results are difficult to understand physically and intuitively. How can both observers measure the other's time to be slower than their own, and yet one observer actually experience less time than another on a round trip? Starting with a constructive model of the physical phenomenon, we will provide a satisfactory explanation for these types of results. With a constructive physical model, we will understand what is really going on. We will be able to see why these results make sense, and we will be able to see why the historical development from the Einstein's formal principle has made these results seem baffling.

So, let's start to construct a model that's simple to think about that represents the basic elements of the physical aspect of relativity. Remember, the basic premises which I am asking you to accept are:

1. Energy always moves at the speed of light through space, and
2. Matter is energy that is trapped in a persistent local repeating (cyclic) pattern.

After we have looked at a variety of patterns of cyclic motion, I think you will be convinced that the same effects will occur with ANY cyclic pattern of energy motion.

FIRST STEP - PERPENDICULAR PLANE: HELICAL MOTION

We can bring these ideas to a scale of speed and size that are easier to imagine. As a metaphor, suppose that a bug represents energy, and the air it flies through represents space. Let's then say that this bug always moves through its local air at the same speed. If the bug flies ahead through air freely, it represents what we commonly call energy traveling through space. But if the bug flies in a pattern that repeats in some cyclic way, it metaphorically represents a particle of matter. Remember, the bug is not the particle, the pattern is the particle, and the bug represents the energy moving within the pattern. Let's focus on the trapped bug pattern as a particle of matter and see what happens to it. For a start, imagine that the bug circles a central light, somewhat like the bugs that endlessly circle streetlights on a summer night. As a first case, suppose that the bug travels in a perfectly vertical circle, in a flat plane, and it always maintains a fixed distance from the light. Every time the bug gets to the top of the circle, we mentally make a ticking sound, and we hear that the ticks are very regular, say one tick per second. We could use the bug pattern as a type of clock.

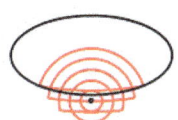
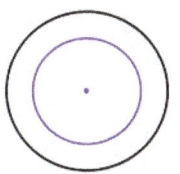

Now, picture what happens if we attach the light that the bug is circling to the tip of the radio antenna of a car, and we imagine that the vertical plane in which the bug is flying is parallel with the front of the car, and therefore perpendicular to the direction of motion of the particle. (We might want to bend the antenna forward a bit so that the bug does not run into it while flying in its vertical plane!)

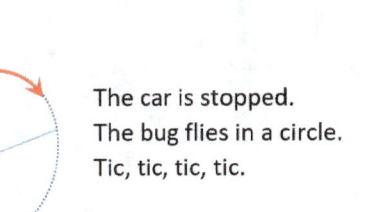

The car is stopped.
The bug flies in a circle.
Tic, tic, tic, tic.

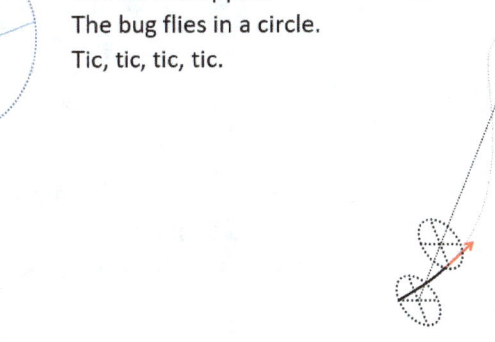

Imagine what happens as we start the car in motion. The bug continues to circle the light, always staying in its plane, and if we are sitting in the car, we still see the bug flying in a perfect circle, because the bug is maintaining the circular pattern relative to the central light. We can still imagine the ticking of a clock each time the bug reaches the top of its circle. But something has changed. The clock is ticking more slowly now because the car is moving.

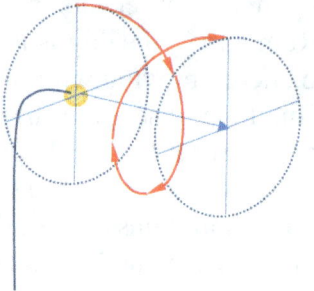

The car is moving.
The bug flies in a helix.
The clock tics more slowly.
Tic...tic...tic.

If we were standing on the street as this car slowly passed by, we would see the bug as travelling in a helical path. Remember that the bug always flies *through the air* with the same airspeed, which we will call 'c' – since that represents the speed of light, the speed that energy moves through space in our constructive physics model. If the car stops, the bug returns to a circular path, for observers both inside the car and for those standing by the street. If the car moves very slowly, the bug flies in a tight helix, barely moving forward with the moving car and the light on the antenna. The difference in the clock ticking rate is barely noticeable.

But as the car gains speed, the helix stretches out, and the bug must devote more and more of his speed c to following the forward motion of car, so the time it takes to make it around this helix to the top position grows longer and longer. As viewed from inside the car, the bug still travels in a perfect circle, but the ticking of the bug clock slows down.

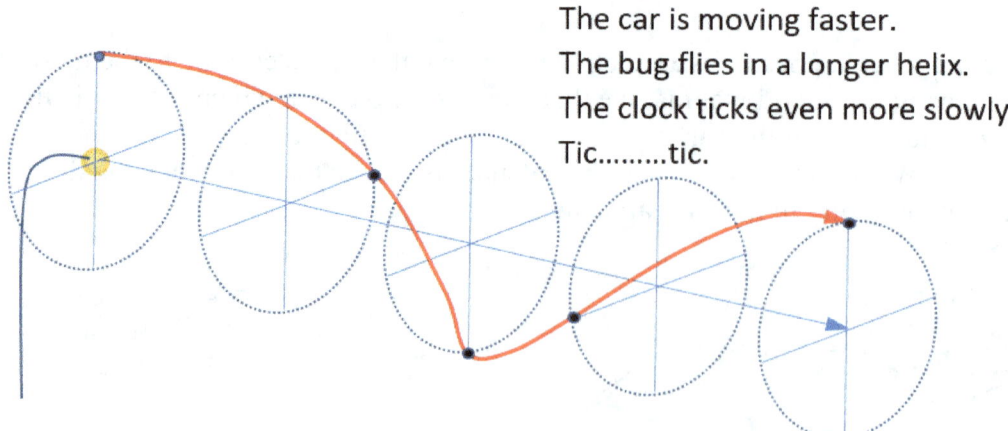

The car is moving faster.
The bug flies in a longer helix.
The clock ticks even more slowly.
Tic.........tic.

As we imagine more complex models like this, we will see that as long as the bug maintains a repeating pattern while moving at a constant speed c through the air, the rate at which the pattern repeats is slowed down. This will happen regardless of the particular shape of the pattern. There will be a couple of other interesting things which happen when other patterns are considered, and we will get to them soon. But we have already discovered one of the essential physical aspects of relativity, namely that when matter (which is energy trapped in a repeating pattern) moves through space, the cycling rate of that energy is slower than the maximum cycling rate it would have if the pattern was not moving through space.

Now I want to make an important point. Although this model of a bug circling a light reminds us of a simple atomic model of an electron circling a nucleus, that image is not exactly what I am trying to imply. What I am suggesting is that this rate of cycling is a rate of TIME as experienced by all matter, a time rate which applies to ALL physical processes in which the matter is involved. We know that most of the mass of an atom is located in the protons and neutrons in its nucleus. But when you get down to the level of the proton and neutron, we again see that internally they are not nuggets of stuff, they are mostly empty space, with a lot of energy moving about inside them. As discussed earlier, if there is nothing but energy inside them (if sub-particles like quarks are also composed of energy), then we expect ALL the inner processes of the protons and neutrons to slow down.

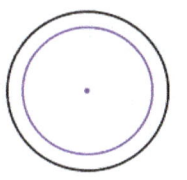

So, when I talk about particles as patterns of energy, and I suggest that the cycling time for these patterns slows down, I'm talking about a slow-down of time on every level. For example, when moving rapidly though space, a non-stable nucleus which has a specific rate of radioactive decay will decay more slowly, and chemical reactions in which structures of atoms are involved will experience a slower rate of chemical reactions, and large living structures will experience a slower rate of neural activity, and slower heartbeats, and slower breathing, and slower aging. Everything about matter that is in motion through space will do what it normally does, but at a slower rate compared to the maximum rate at which it would do the same things if it were not moving through space.

All these rates will be slowed by the same amount, and because of this, a person traveling in a very fast spaceship would experience a kind of 'personal time' which would be slower than the 'personal time' experienced by someone not moving through space as fast. Although there are other details to be examined, this is the physical reality behind the famous 'twin paradox' of relativity. The space-traveling twin comes home younger than the stay-at-home twin because the traveling twin has a slower personal time. And it is equally important to note that the traveling twin would not be able to perceive this slower rate, because everything in his or her moving ship would be running at the same slowed rate: the clocks, the computer circuits, literally everything: absorption of oxygen, synapse firing rate, breathing rate, pulse, blink rate, heat transfer rates, everything.

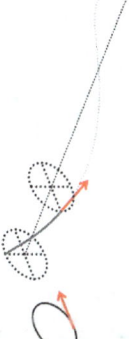

How much is time slowed, numerically? This is quite simple to compute. If we go back to the bug model, we can see that the bug must divide its speed c vectorially into two functions: 1) keeping up with the bulk motion of the car, and 2) circling the light. This vectorial division just means that the speed c forms the hypotenuse of a right triangle, and one side of the triangle represents the forward motion of the car (call it v.) The velocity not dedicated to forward motion is the circling speed, which is the other side of the triangle, namely $\sqrt{(c^2 - v^2)}$ by the Pythagorean Theory. When the speed v is zero, the circling speed is c. Therefore, when the car is moving at speed v, the cycling speed is slowed by the ratio $\frac{\sqrt{(c^2 - v^2)}}{c}$. As a simple example which I will often use, if the car is moving at (.866)*c, (v=.866c) the cycling rate is ½ of the normal rate. This is because .866c squared is .75c², and c²-.75c² = ¼ c², and the square root of (¼) c²= (½) c. To be consistent with the language of relativity, we would say that in this case, gamma = 2. Gamma is the key parameter in relativity that defines how strong the "relativistic effects" are in any situation. If gamma = 1, then there are no relativistic effects. Gamma can only be one or greater. Gamma is defined as:

$$\text{gamma} = \frac{c}{\sqrt{c^2-v^2}} = \frac{1}{\sqrt{1-\frac{v^2}{c^2}}}$$

That is, if moving through space at 86.6% of the speed of light, the cycle takes twice as much time to complete as it would take to complete the cycle if not moving through space. Looking down on two circles (almost, but not quite straight down, so we can still make out the circles,) we see the relevant triangle for these two cases in the pictures below. At left, if the velocity of the pattern is zero, all the speed c of the bug is in the plane of the circle, but at right, if the bug has to keep up with the car that is moving forward at .866c, then the total speed c has to be partly forward (.866c) and partly in the plane of the circle (.5c) These two vectors add up (with the Pythagorean Theorem) to the vector c which is the speed of the bug through the air.

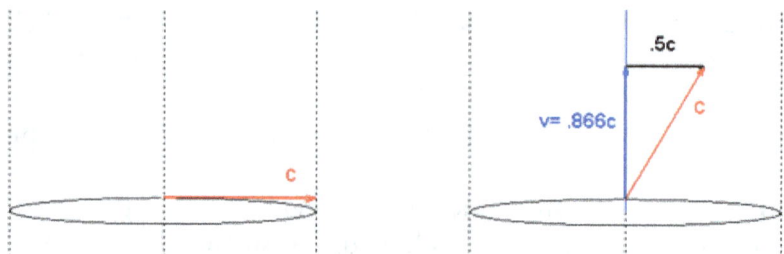

In our model, if the car speed is 86.6 % as fast as the speed of the bug's flight speed through the air, the 'bug clock' will tick twice as slowly. If, when the car is stopped, the bug takes one second to make a cycle flying at speed c through the air, then when the car is moving forward at .866c it will take two seconds for the bug to complete the cycle. What happens as the car goes faster? At v = .9428c, the bug clock would click at 1/3 the normal rate. At higher speeds, the bug's helix will stretch out longer and longer, until, if the car goes at speed c, the bug will have to devote ALL of its flight speed to keeping up with the car. The bug will be flying straight ahead, and not cycling at all. The bug clock will be stopped. For real particles in physics experiments, it takes enormous amount of energy to accelerate matter to speeds near the speed of light c. The cycling time can be greatly slowed for tiny particles in accelerators, but it cannot be stopped. The physicists might say that it would take infinite energy to make that happen. But just coming very close takes a whole lot of energy.

I wrote a computer program to demonstrate this motion, and a snapshot is shown below. What cannot be seen in the snapshot is that the 'bug' on the left travels TWICE around its stationary circle as the bug on the right travels ONCE around its moving circle (helix,) because the forward speed of the circle is v =.866c, so that the gamma ratio is 2. Put another way, the path distance along the helix is the same distance as two trips around the circle.

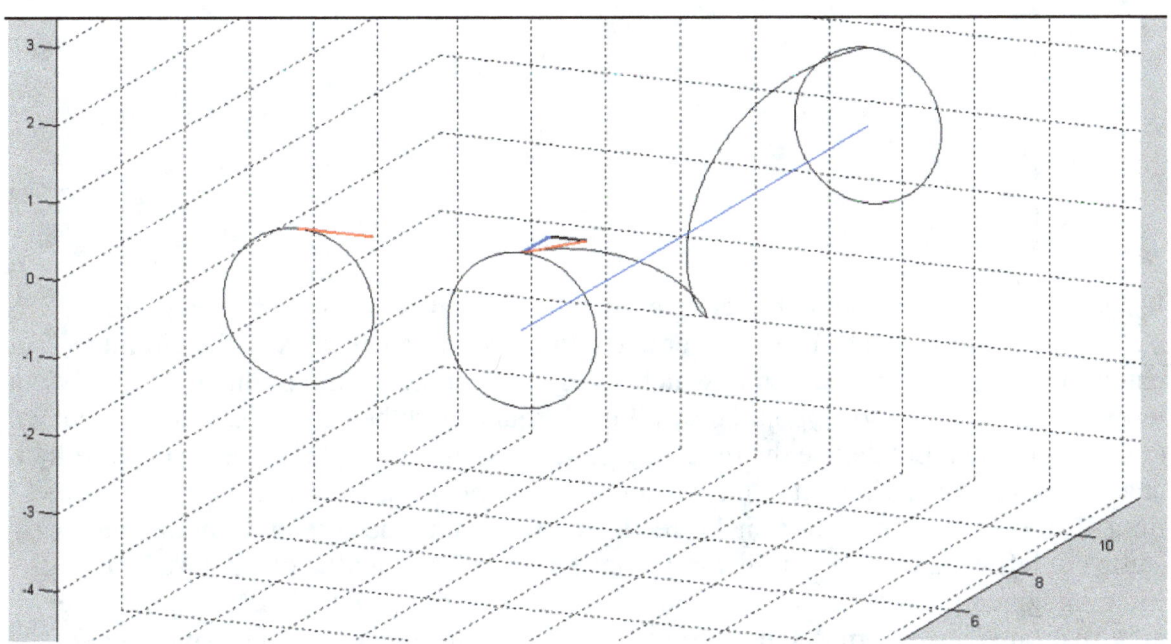

SECOND STEP - PARALLEL PLANE RADIAL MOTION

Now let's take the next step toward developing a constructive model of relativity. But before we do, I want to emphasize again that we are building a relativity model from the bottom up. I have not yet talked about any observer's measurements; in particular it is clear that I have not explained how two different observers moving relative to each other can both measure each other's time as going slower than their own time. The idea here is to start with an explanation of how the phenomena works physically, and THEN explain how the observer's measurements turn out the way they do. This is quite a different approach than that taken by Einstein, and almost every other relativity book you will find.

Now, as a second step, let's look at a different situation. In the first step we imagined a repeating energy pattern that took place entirely in a plane perpendicular to the direction of motion of the pattern. Now, let's consider an energy pattern that takes place in a plane which is parallel to the bulk motion of the pattern. Returning to the car, let's consider a plane which is horizontal. We are going to spend a significant amount of time considering the motion of energy, represented by various bugs traveling in this plane, and in so doing, the fundamentals of relativity will become clearer. So, let's start as simply as we can, and build the ideas gradually.

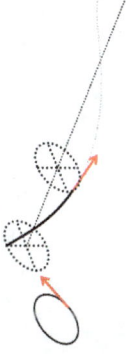

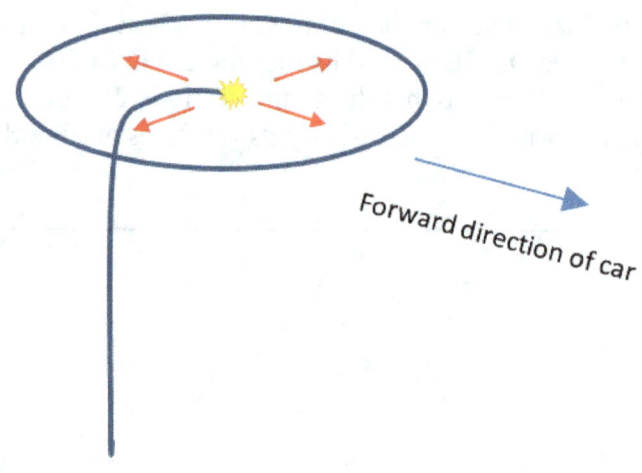

Instead of starting with a bug traveling around the central point in a circle as we did before, let's start with a different pattern, and then we can come back to the circling pattern later. Consider a pattern in which 4 bugs start at the central point, and they fly out in four directions. At some point, they all turn around, and they come back to the center at the same time. Let that be the repeating pattern for this example. After that, we'll look at a swarm of bugs flying out in all directions in this plane, and all returning at the same time. But first, let's focus just on 4 directions. Relative to the car, let's say that one bug starts out leftward, one starts rightward, one starts forward, and the other starts rearward. First, imagine the car is sitting still, and the bugs all fly out an equal distance, then they all turn around and return to the central point. They must meet up at the same time, of course, so that their pattern of motion is the same on each cycle. After one complete cycle, they turn around and fly outward again, and then inward to come together again… out and in as a repeating pattern. Again, this bug model is intended as a metaphor for energy which is trapped in a repeating local pattern. Because the bugs always travel at the same speed through the air, they represent energy which always travels through space at the speed of light. Our goal is to understand what happens when this pattern itself is moving through space.

First consider the two bugs which are going laterally, that is, to the left and right. What happens to them when the car is put in motion? They find themselves in a situation which is very much like the lone bug we considered earlier in the first model. To maintain their position in the pattern, they must devote some of their speed c to keeping up with the motion of the car, and another part of their motion to fulfilling their lateral cyclic pattern. Again, their speed is split vectorially, and the right triangle which represents this vectorial split has a hypotenuse c (flight speed), a second side v (car speed) and a third side that represents their rate of accomplishing the cyclic pattern. Just as in the first model, as the car speeds up, these laterally moving bugs take more time to complete the round trip. Just as in the first model, if the car moves at v=.866c, gamma=2, that is, the cycle takes twice as much time to complete as when v=0 and the car is stopped.

So far so good, but now we consider the interesting situation involving the other two bugs in the pattern, the ones flying forward and rearward. What happens to them? When the car is in motion the bug moving forward at speed c through the air finds that his turn-around point moves away from him as he flies toward it. As far as his progress toward completing the cycle, it takes him much more time to reach his forward turnaround point than it takes him to get back. Note that he doesn't really feel that he's going 'upwind', and then 'downwind', because his speed through the air is always the same. Clearly, though, the two bugs which either start out going forward or start out going rearward will take the same time to complete their round-trips, their long trips and short trips are just in opposite sequences.

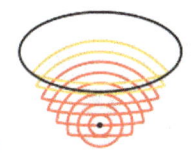

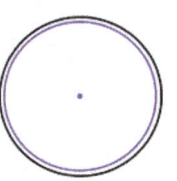

We can use the same sort of vector reasoning to compute these forward & rearward bugs' speeds relative to the car and their travel times, but in their cases, the 'triangles' collapse to lines, because v and c are aligned with each other. So, the analysis is actually simpler here. When moving forward, these bugs' speed (relative to the antenna) is c-v, and when moving rearward, their speed (relative to the antenna) is c+v. Now here's the really interesting part. If we imagine that these two bugs fly out from the central point equally as far forward and rearward as the left-right bugs move laterally, and we total up how much time it would take them to complete the forward-rearward cycle, we will see that it would take the forward-rearward bugs **more** total time to complete their cycle than it takes the left-right bug to compete their cycle.

When the car is moving, in order for all four bugs to maintain their mutual cyclic pattern of *returning at the same time*, pulsing in and out from the central point in a harmonious repeating cycle, the forward-rearward bugs *must* travel a <u>shorter</u> distance away from the antenna than the lateral moving bugs travel. How much shorter must the longitudinal bugs' travel distance be so that all 4 bugs return at the same time? The math is quite simple and is shown below. We compute their total flight time as a function of the unknown distance d which the forward-rearward bugs must travel. We set their total cycle time equal to the flight time of the left-right bugs, and solve for the unknown shortened distance d. If we let capital D represent the turn-around distance the left-right bugs, we find that the ratio D/d turns out to be exactly equal to gamma! In other words, the shrinkage of the forward-rearward distance required for a harmonious cyclical return is numerically equal to the amount that the cycle is slowed down in time. Using the same car speed example as before, if the car is moving at 86.6% of the bugs' flight speed, (v=.866c) and if the forward-rearward bugs go half as far away from the central point as the left-right bugs, all four bugs will return to center at the same time, and the cycle time for all 4 bugs will be twice as long as it would be if the car were stopped.

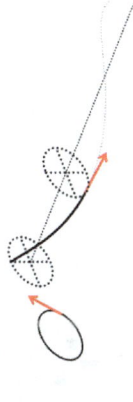

4 bug Velocity Diagram (Direction of motion of car = up)

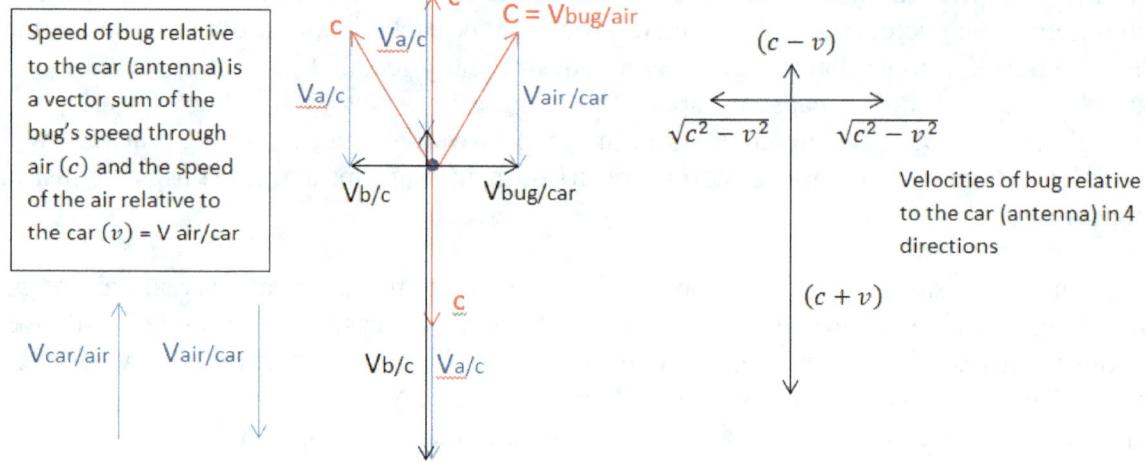

Speed of bug relative to the car (antenna) is a vector sum of the bug's speed through air (c) and the speed of the air relative to the car (v) = V air/car

Velocities of bug relative to the car (antenna) in 4 directions

Solution finding D/d that gives a common return time for the 4 bugs

$$\text{speed} = \frac{distance}{time}; \quad \text{time} = \frac{distance}{speed}$$

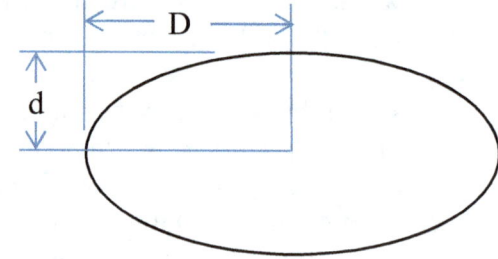

D = lateral distance; d = longitudinal distance

Lateral speed relative to pattern center = $\sqrt{c^2 - v^2}$

Longitudinal speed forward relative to pattern center = $c - v$

Longitudinal speed rearward relative to pattern center = $c + v$

Lateral cycle time = $\frac{2D}{\sqrt{c^2-v^2}}$; Longitudinal cycle time = $\frac{d}{(c-v)} + \frac{d}{(c+v)}$

Setting these two times equal,

$$\frac{2D}{\sqrt{c^2 - v^2}} = d \left\{ \frac{c+v+c-v}{(c-v)(c+v)} \right\}$$

$$\frac{2D}{\sqrt{c^2 - v^2}} = d \left\{ \frac{2c}{(c^2 - v^2)} \right\}$$

$$D = d \left\{ \frac{c}{\sqrt{c^2 - v^2}} \right\}$$

$$\frac{D}{d} = \left\{\frac{c}{\sqrt{c^2 - v^2}}\right\} = \frac{1}{\sqrt{1 - \frac{v^2}{c^2}}} = gamma$$

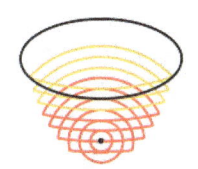

We will spend more time examining bugs flying in this plane, but we can already begin to grasp the second fundamental aspect of relativity. The standard development of relativity from Einstein's formal principle regularly mentions two aspects of relativity: time dilation and length contraction. The standard relativity books typically talk about these as aspects of the measurements of objects that are moving relative to the observer's own frame of reference. Time dilation is the observed 'slowing down of time for the observed object', and length contraction is an observed 'shortening of the length of observed objects in their direction of motion.' The magnitude of both these effects are equal to gamma.

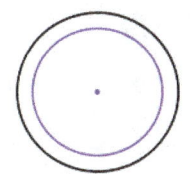

If we consider all particles and sub-particles of matter to be energy trapped in cyclic local patterns, then from the 4-bug example, we can start to see that material structures that are built from these particles might have a shortened length as well as a slowed atomic and chemical time rate when the entire pattern is moving through space. This shortening in the direction of motion would not involve any physical pressure, stain, or stress; it would simply be that the energy patterns which are normally circular or spherical when not moving through space would be elliptical or flattened spheroids when they are moving through space in order to maintain their harmonic cyclical energy patterns. In the next section, I will show that by adding more bugs to the model we really do get an elliptical shape. Then we will get into the third very important aspect of relativity which is not as well-known as time dilation and length contraction. It is this third aspect which will open the door to a full understanding of many of the well-known paradoxes involving the symmetry of relativistic measurements.

"We know that these reasonings do not come from us and at the same time, we recognize in them, because of their harmony, the work of reasonable beings like ourselves. It is this harmony, this quality if you will, that is the sole basis for the only reality we will ever know."

- Robert Pirsig, Zen and the Art of Motorcycle Maintenance.

Now let's look more closely at the situation of the harmonic energy pattern in the horizontal plane. Rather than just four bugs, let's expand the idea and have a vast number of bugs all participating in this cyclical motion. Again, we will have them start at a central point of the pattern, and fly outward, but now instead of just four directions, we will have hundreds of bugs, and we'll have them fly out from the center in all possible directions in the horizontal plane. Naturally, they will form an expanding circle. If the car is not moving the bugs simply fly out to the perimeter of a circle then they will all turn around at the same time, and they all return to center at the same time. This represents matter which has no motion through space as energy pulsing outward and returning to the center in unison repeatedly.

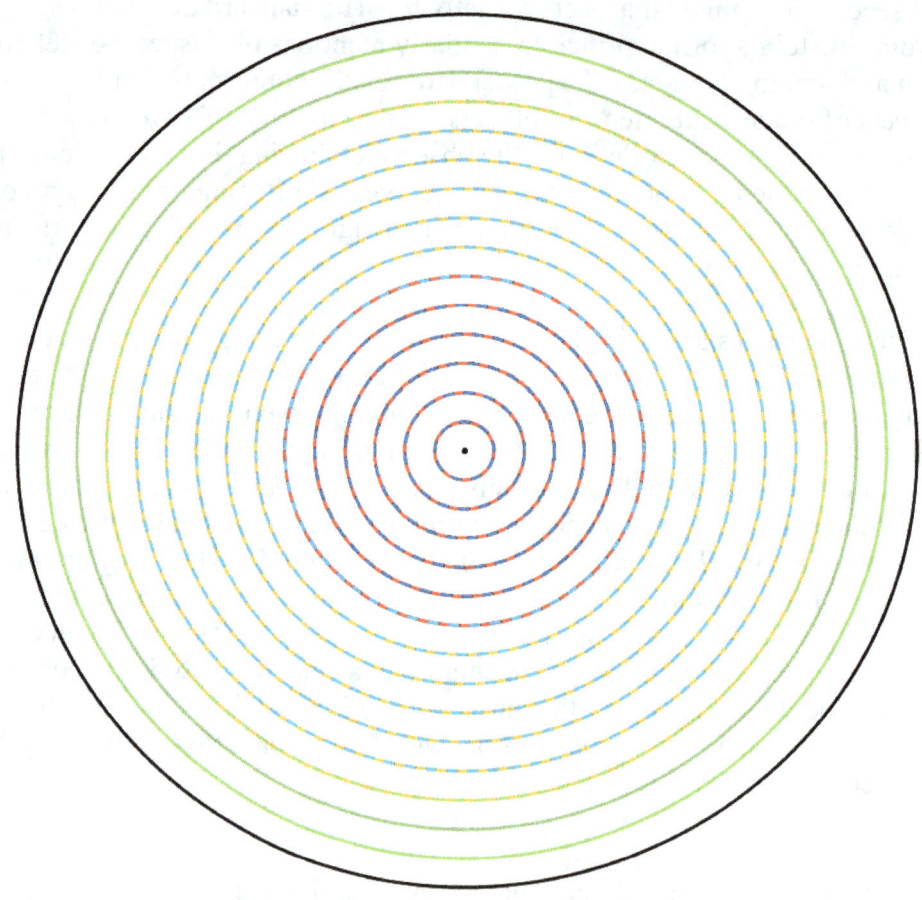

Bug Swarm Plot #1

The graphic shows the positions of the bugs at various times when the car is not moving. The bugs in their cyclic pattern, again, metaphorically represent energy moving within a particle of mass. The energy always travels at the speed of light through space, just as the bugs always travel at the same speed through the air. When the central point of the pattern is not moving, the bugs spread out, and then return, always in perfect circles. The pattern reminds us of ripples in a pond, but in this case, there is a ring which reflects the waves back to the center, where they all meet up at the same time so that the pattern can begin again harmoniously.

The positions of the bugs when the they are moving outward within this pattern are shown in red, yellow, and some green, and their positions while returning are shown in green, light blue, and dark blue. Because their outward and inward positions are so symmetrical, the light blue and dark blue rings for the return positions in the diagram virtually overlay the red and yellow rings that traced their outward motion. Dotted lines were used here so that both colors can be seen.

Now, let's look at the picture of this wave of bugs when the car is moving, and for this next picture, imagine that <u>the car is moving toward the right.</u> For this next graphic, I have chosen a speed v=.5c, which corresponds to a gamma of 1.15. In other words, the car speed is half of the bug's flight speed through the air, which produces only about a 15% slow-down of the cycle time relative to the cycle time when the car is not moving. I think this makes things a little easier to see for the initial discussion, then I will also show our more familiar case of v=.866c later.

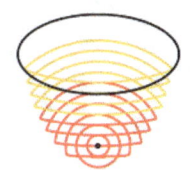

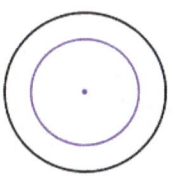

For this next picture, imagine that a camera has been set up on a tripod which is attached to the car, and the camera is positioned directly above the tip of the antenna. So, although the car, and the antenna, and the bug pattern as a whole are all moving to the right through the air, the camera has a fixed position relative to the central tip of the antenna. The camera moves along with the car.

Now imagine that this camera has strobe lights that only illuminate the bugs when the light flashes. The shutter is left open so that every flash shows a new ring image of where the bugs are at each flash, and all the rings are captured on one photograph. Imagine further that the strobe light's timer has been set to flash exactly 30 equally spaced times during one complete bug cycle, and that a series of colored filters are used, and these colors change every 6 flashes. The following picture is what the camera would capture.

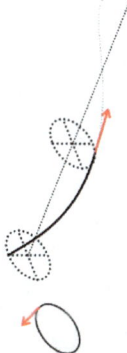

33

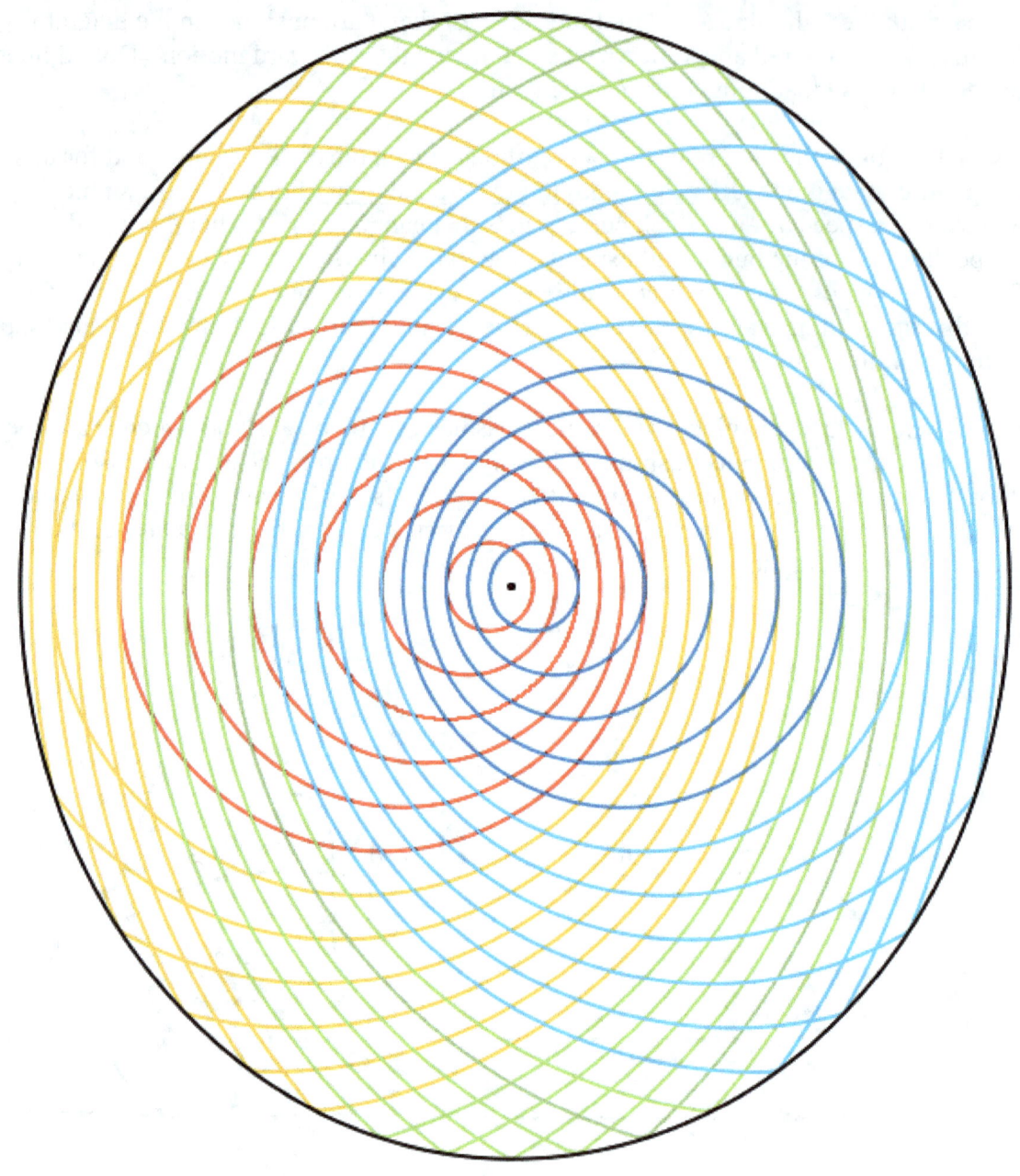

Bug Swarm Plot #2

Study this picture carefully as I point out the many interesting features of our constructive relativity model which it represents. First, you see at the center a tiny black dot, which is the antenna, the central point of the energy pattern. This never moves in the photograph, because the camera is attached to the car and travels with it. Each colored ring represents the position of our swarm of flying bugs as they fly out from, and then in toward the central point of the pattern. This is the same scheme that was used in the first case in which the car was not moving. But where in that first case, we saw evenly spaced circular rings, here we see a much more complex pattern. The rings of bugs are in fact still circular, because they all travel at the same speed through the air, but these circles are no longer concentric.

Each ring represents the position of the bugs at one point in time – the flashing strobe light has colored these rings to make it simpler to follow the sequence. The color sequence follows the familiar 'colors of the rainbow' sequence called ROY G BIV – for Red, Orange, Yellow, Green, Indigo, Blue and Violet – but I only used 5 colors: Red, Yellow, Green, Light Blue and Dark Blue. So, in the picture, the red rings represent the first 6 tics of time, the yellow rings represent the next 6 tics of time, and the last 6 dark blue rings represent the positions of the bugs at the last 6 tics as they converge to the central point of the pattern. Unlike the first case in which the car was not moving, now we can see all the colors.

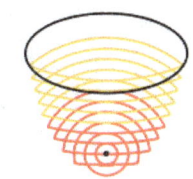

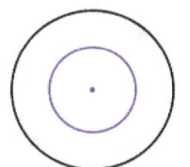

The graphic presented is a picture of the local pattern in the reference frame of the moving car – this is very important. What I mean by this is that the central point of the pattern is always kept at the center: it does not change position because in the picture presented we are traveling along with the pattern as we view the bugs' positions. Remembering that the car is moving to the right, notice that in the first 'red' time period, the rightmost 'forward-moving' bugs are making slow progress toward the perimeter, but the leftmost 'rearward-moving' bugs get more than halfway to the perimeter by the end of the first six red clock ticks. During the second, 'yellow' time period, the rearward moving bugs are turning around, while the forward moving bugs are only halfway to their turn-around perimeter.

Study the picture some more. Notice that the lateral moving bugs which started out perpendicular to the bulk rightward motion of the car make their turn-around right in the middle of the green set of clock ticks. Their speed within the pattern, represented by the spacing of the green rings is equal on the outward and inward parts of their cycle.

Notice one more very important thing. The shape of the perimeter is an ellipse. Since the motion of the pattern is to the right, the perimeter which was a circle in the first ring diagram has been compressed in the direction of motion. If the perimeter was not an ellipse of just the right shape, the bugs would not all return to the central point at the same time, and the pattern would not therefore be a harmonically repeatable pattern. And, as I described earlier in the four-bug example, the amount that the ellipse is 'compressed' i.e. the ratio of the vertical to the horizontal dimensions of this ellipse is equal to gamma – which is numerically the same as the amount of the time-slowing which occurs relative to pattern's motion when the car is not moving.

You might reasonably have a few questions about how this picture was created, and I hope to explain all these questions now. The graphic was created, of course, with the help of a computer program. The position of the bugs at each instant of time was computed just as I explained in the four-bug example, by a vector analysis of the velocity of each bug's flight with the following conceptual relationship: the velocity of the bug relative to the pattern is equal to the velocity of the bug's flight speed through the air (c)

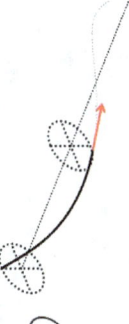

minus the velocity of the pattern's motion through the air (v.) Note that this is a vectorial subtraction of velocities, not a linear one. In the four-bug model, we were able to use a right triangle for the lateral moving bugs, and we used simply (c-v) and (c+v) for relative speeds of the forward and rearward moving bugs, respectively. For bugs moving in the other directions, this vector subtraction of velocities is slightly more complicated, but still the simple rules of trigonometry do the job. The triangles involved are not right triangles, but these triangles can be solved with the law of sines.

Using these 'velocities relative to the pattern' for each bug, the positions of each bug at each time step were plotted, and the bugs were allowed to progress on their courses until they hit the elliptical perimeter, then they were turned around, and similar calculations plotted their return courses. I ran this program for various values of v/c, and I have shown you two of them. The first was for the still car, which gave the circular pattern. For the first one, v=0 and gamma=1, in other words, there was no cycle time slow down, and the ellipse was actually a circle. The second picture which we have been discussing used v/c = .5, (car traveling at half of bug's speed through air) which gives a gamma=1.15, where the cycle time takes about 15% longer, and the ellipse is squashed about 15% The following picture is for the case v=.866c, which produces a gamma=2, in which the cycle time is twice as long as the cycle time for the v=0 case, and the ellipse is squashed into a 2:1 shape. In this picture, the relativistic effects are stronger than in the previous picture.

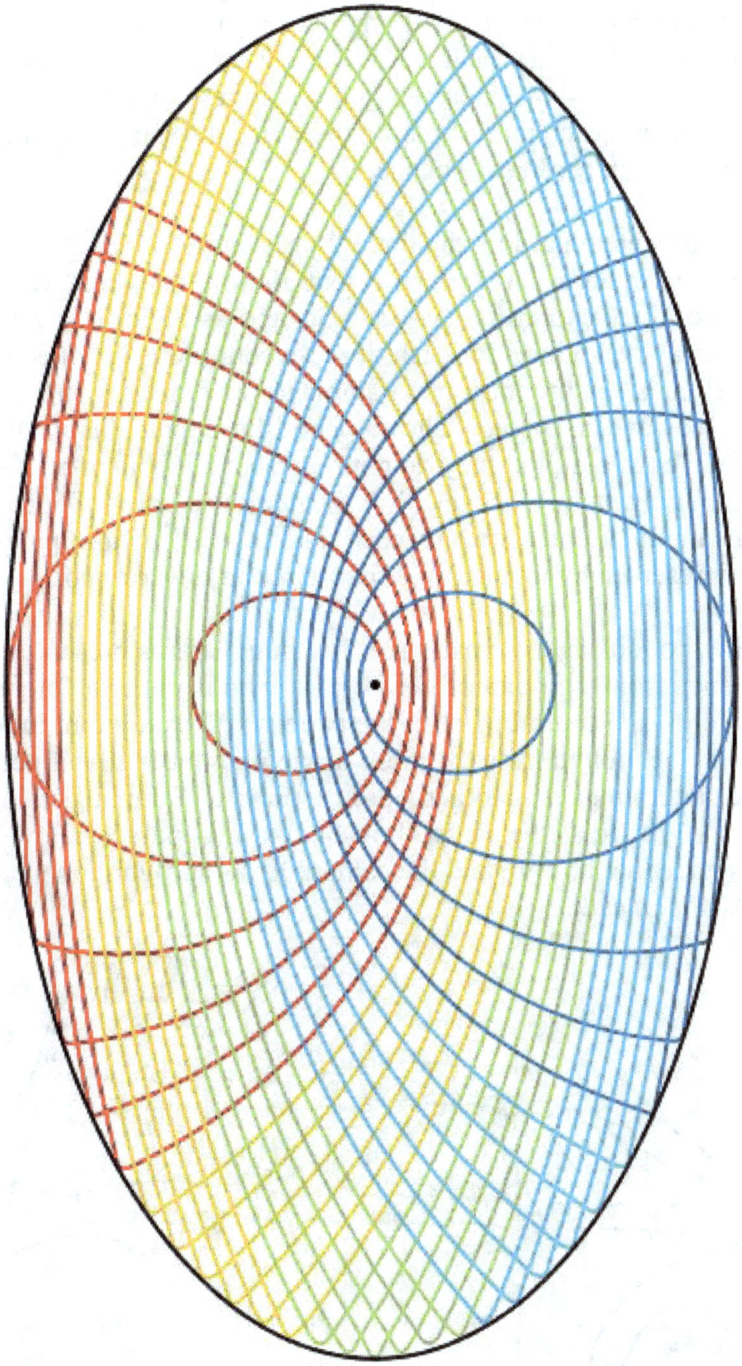

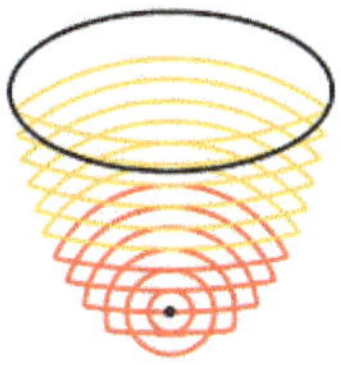

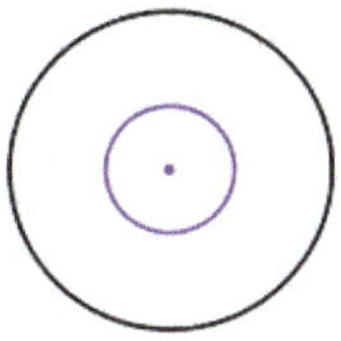

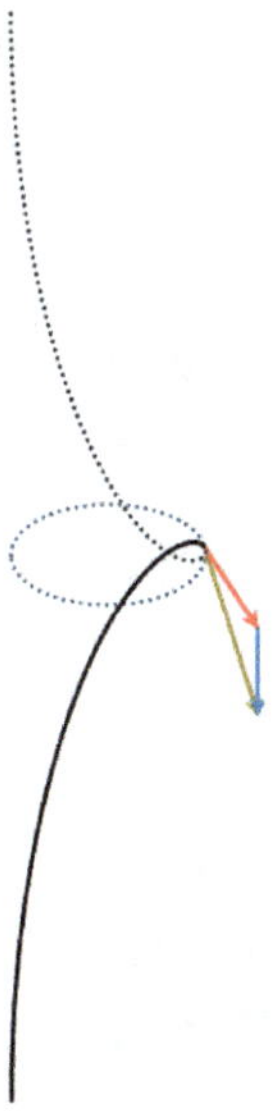

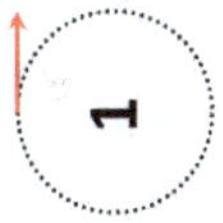

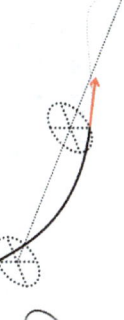

Bug Swarm Plot #3

Notice that in this v=.866c, gamma = 2 picture, there are again 5 colors, with 6 time tics for each color, so a total of 30 tics of time. In this case, the forward (right) moving bugs spend 28 of these tics flying forward toward the perimeter, and make their return flight in only two tics, but the bugs that start by flying rearward turn around after only two of the

30 tics, and spend most of their flight time (28 tics) flying forward, returning to the center of the pattern. Note again that the lateral-flying bugs turn around right in the middle of the central green time band, 15 tics each way. There is an obvious symmetry here, because no matter what path is taken, each bug has another bug which flies in the opposite direction to it, and each pair of bugs experience the same trip time, completing similar legs of their cycle in reverse sequence.

I need to add one thing here regarding the three 'bug-swarm' or energy ring plots that I've just presented. I used the 5 colors to represent 30 time steps required to complete a whole cycle at each of three conditions (v=0, v=.5c, and v=.866c) It is very important to emphasize that the total cycle times for each of these conditions is not equal, rather the total cycle times get longer in proportion to gamma as the pattern's bulk speed through space increases, as we have discussed. The colors of these plots therefore represent the phases of the cycle in each plot, but they do not represent equivalent time steps from picture to picture. For example, since the 3rd 'swarm plot' is for v=.866c, where gamma=2, the time interval represented by each tick in the 3rd swarm plot equals twice the time interval represented by a tick in the first (circular) swarm plot.

There is another way to look at these patterns. The bug patterns presented above show the motion of the bugs from the point of view of one who is traveling along with the pattern. In other words, the central point of the pattern is a fixed point in the plots. This same pattern can also be plotted from the point of view of one who is fixed in the space through which the pattern moves, in other words, a picture of the bugs taken by a camera that is fastened to the ground as the car, with its bug swarm, passes below it. Imagine that the car passes under a platform, and the camera looks straight down to record the bug pattern as it passes underneath.

If plotted this way, we get a different picture which is also very interesting. For this situation, I'll show the development of the pattern with a series of snapshots.

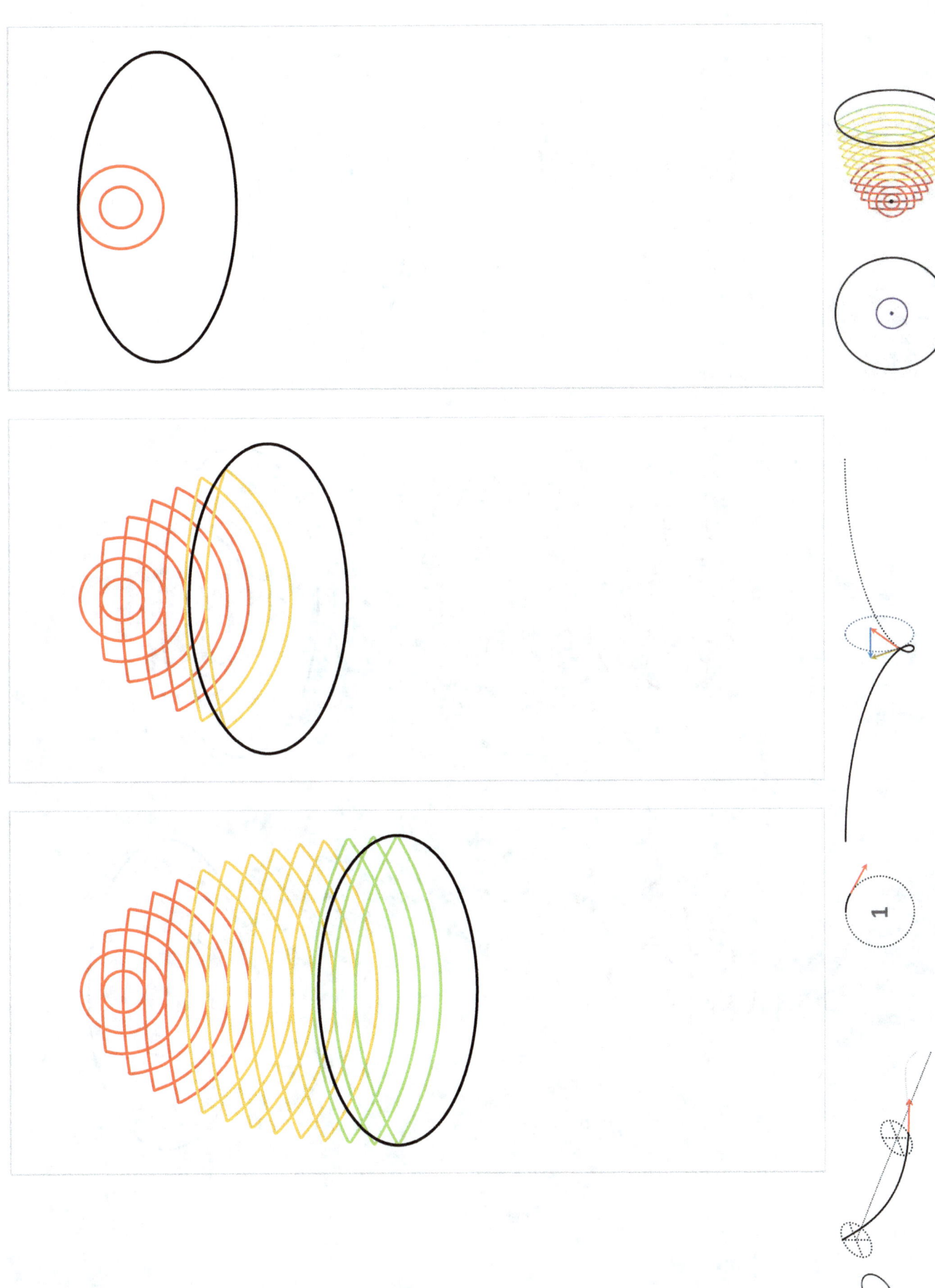

39

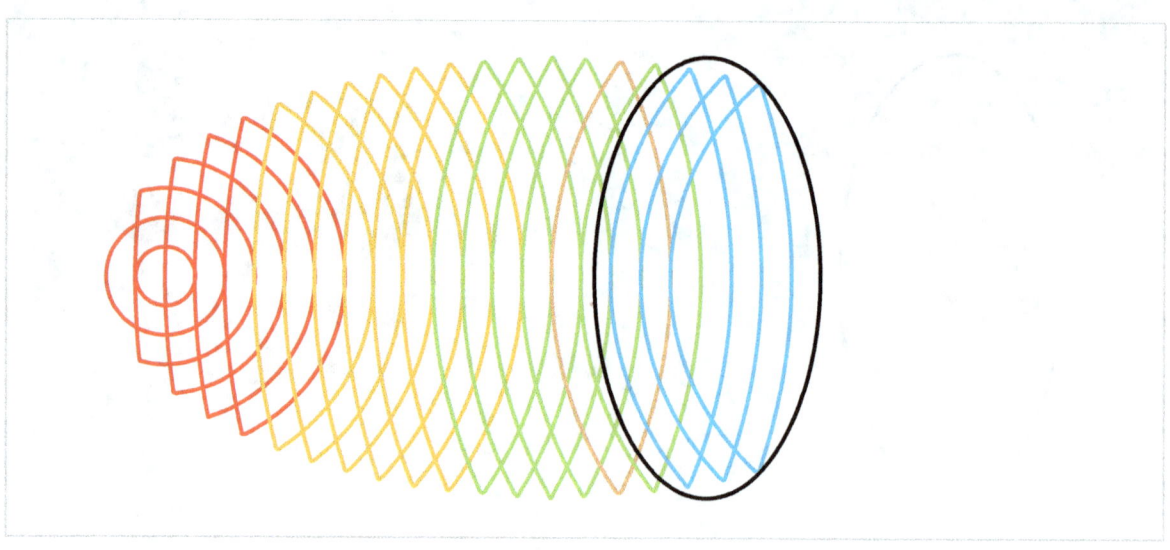
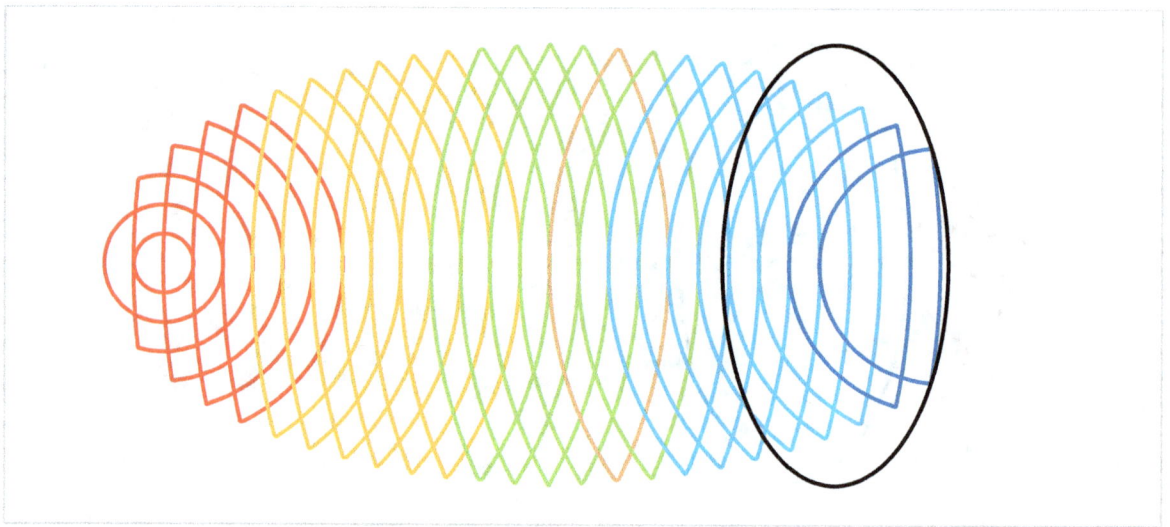
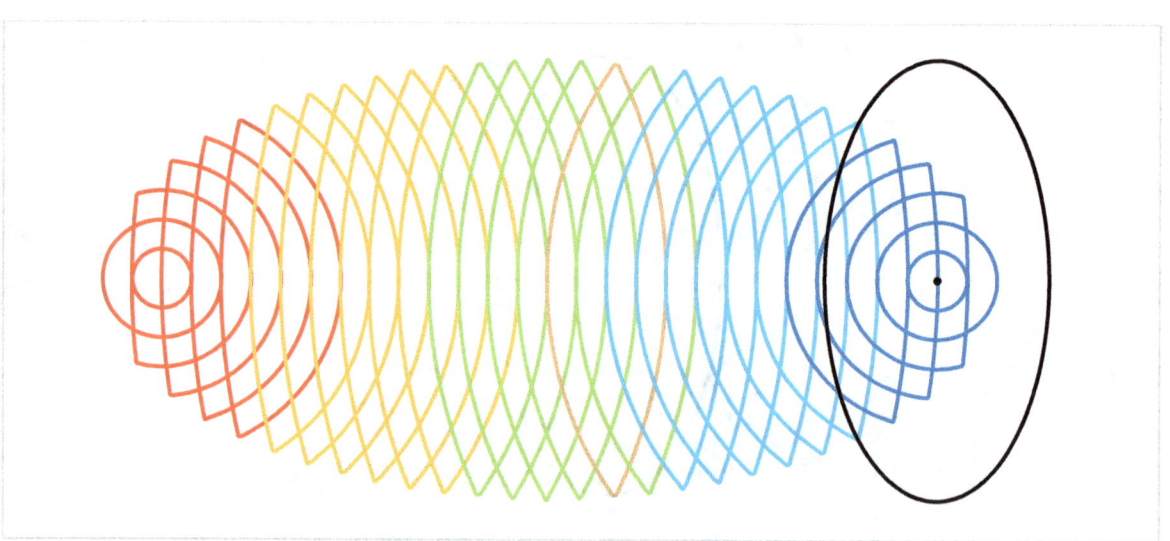

In this series of plots, which assumed a viewpoint which is fixed in space, we can see that the bugs indeed all fly at the same speed through space, because the spacing of the rings is uniform, and the rings, as viewed in the reference frame of the stationary air, are circular and concentric. We also see that all the bugs return to the center of the pattern at the same time when their turn-around boundary is an ellipse whose height ratio is gamma.

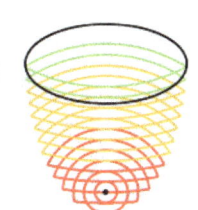

To show that this ellipse boundary is critical, look what happens when the boundary is a circle instead of an ellipse:

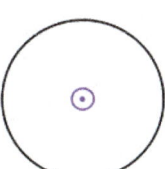

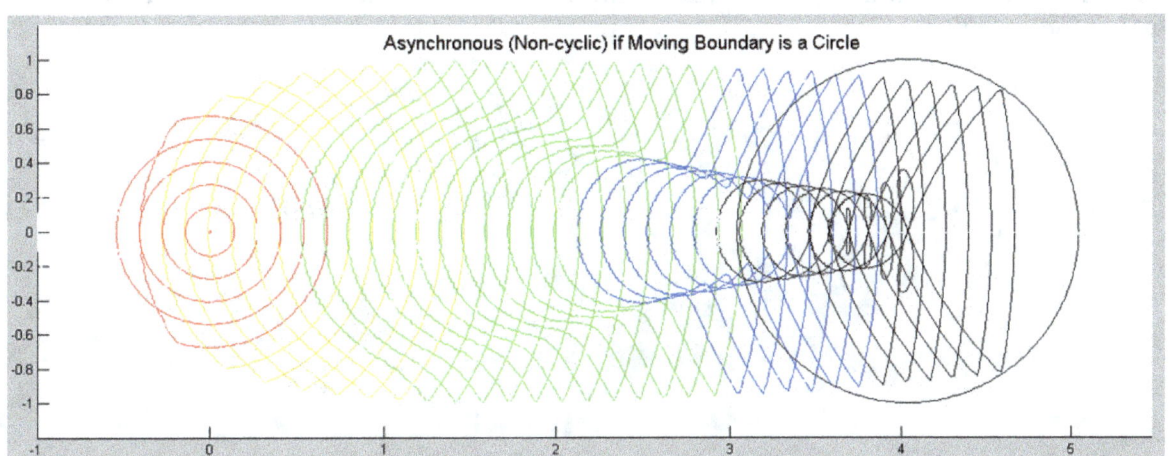

In this picture, we see that the bugs would not harmoniously return to the central point. If the turn-around boundary was a circle, the bugs would not return at the same time, and the pattern would be lost. But with the correct (shortened by ratio gamma) elliptical boundary, the bugs return at the same time, and are ready to repeat the pattern in another cycle.

If we use our imagination a little, we can now picture spherical swarms of bug radiating in all directions from a central point, (not just in a horizontal plane,) flying to a periphery, and returning to the center in a pulsing, repeating pattern. We should also be rather confident that if each individual bug always flies at a constant speed c through the air, then if the central point they are pulsing about is moving through the air at some speed v which is slower than c, the shape of the swarm will be a flattened sphere (ellipsoid), flattened by the ratio gamma in the direction of motion of the pattern as a whole. And we can also expect that the pulse rate of this pattern will be slower (by the factor 1/gamma) than the pulse rate when the pattern is not moving.

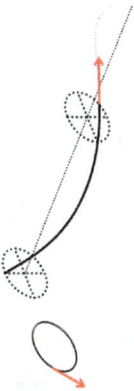

THIRD STEP – PARALLEL PLANE CIRCULAR MOTION

To further convince you that ANY cyclic motion within a moving pattern would produce the same slow-down in cycle rates and length shrinkage, I'll show another graphic. We started the whole bug metaphor constructive model argument with a bug flying around a circle which was perpendicular to the direction of motion of the pattern – the helical path. Now let's look at motion of a bug flying around a circle in a plane which is parallel to the direction of motion. There won't be any helix here, as the bug will always be in one plane, whether observed from within the car, or observed by someone standing on the side of the road. Will this sort of cyclical motion be consistent with the other motions we've looked at?

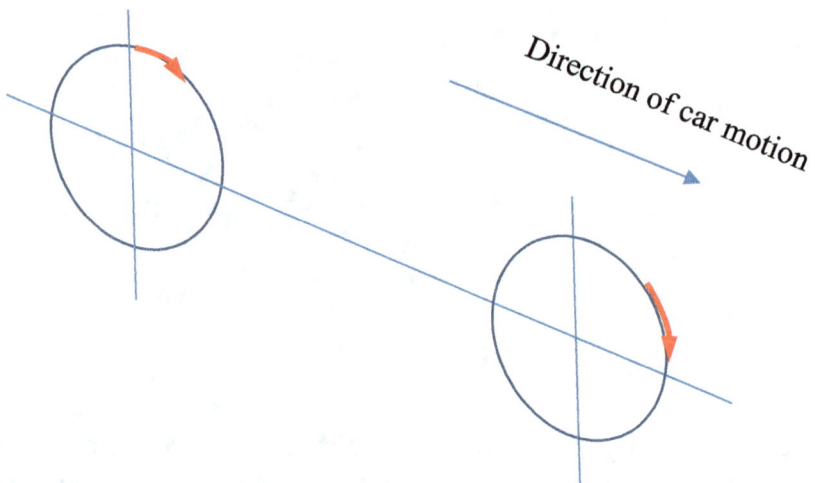

Again, the bug must travel at a constant speed through air, as a metaphor for energy traveling at a constant speed through space. And again, in exploring this model, we find that we must send the bug around an elliptical path of height ratio gamma in order to obtain the same slow-down which we have observed in the other cases. For example, if we started with two bugs at the highest position, and sent one bug in the transverse plane around a circle (helical path as viewed from the ground) and the other bug in this longitudinal plane (aligned with the direction of motion,) in order for them to re-unite after one cycle, the 'circle' in the longitudinal plane would need to be flattened into an ellipse by the gamma ratio, as shown in the graphic below. These snapshots represent observations by an observer on the ground.

Displayed in the graphic are the path of the bug at speed c (red arrow,) the speed of the bug along its tangential path around the ellipse (brown arrow,) and the velocity of the air relative to the pattern (blue arrow.) Note that the velocity of the air relative to the pattern is simply the negative of the velocity of the pattern through the air. As before, the velocity c must be vectorially split between progress around the ellipse, and v dedicated to keeping up with the motion of the pattern.

$v_{tangent} + v_{pattern} = c$ or, as shown in the vector diagrams, $v_{tangent} = c - v_{pattern}$

Trigonometry (the law of sines) was again used to solve for the vector angles required to keep the bug on the ellipse as the ellipse moves through space at speed v.

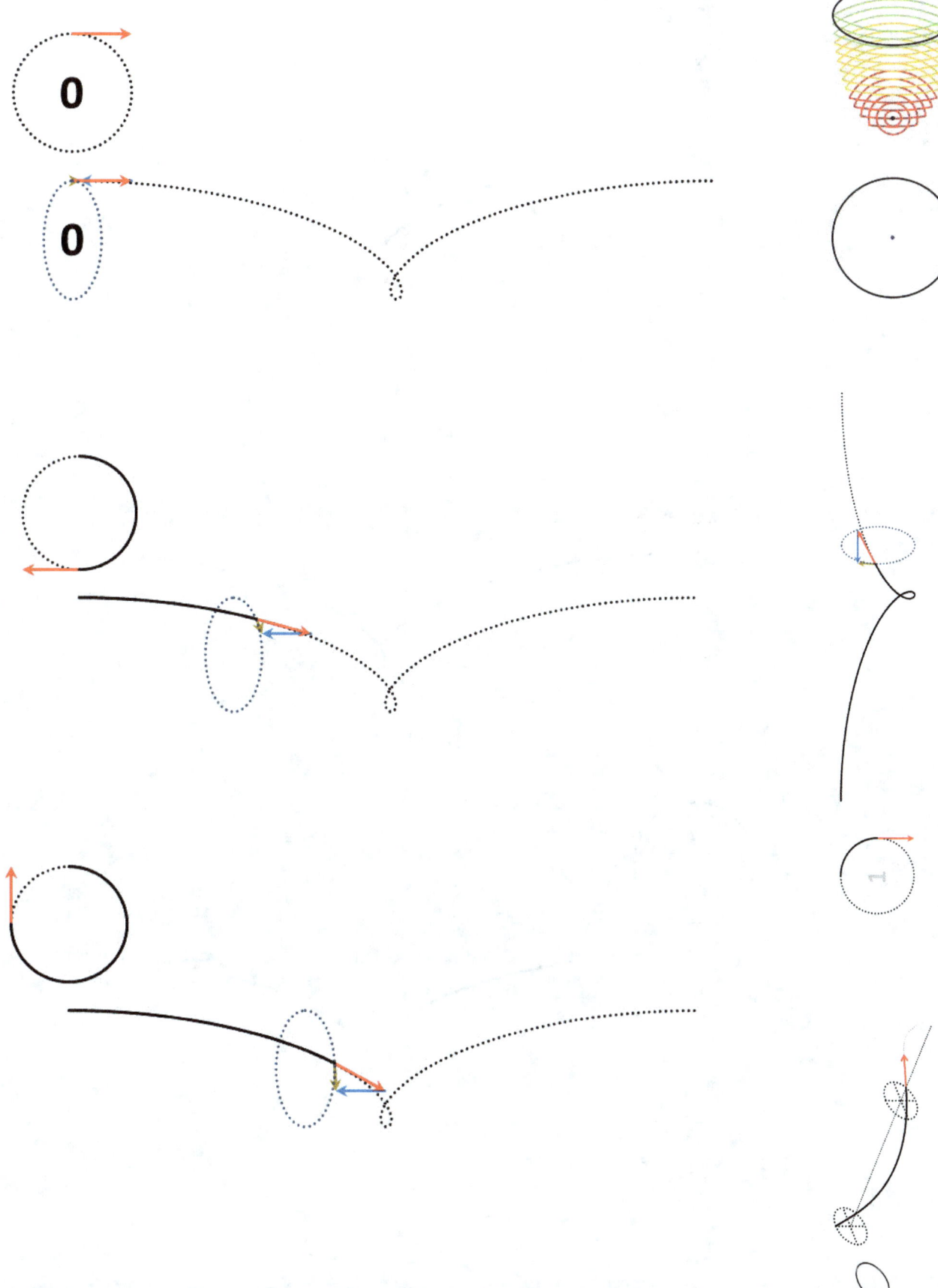

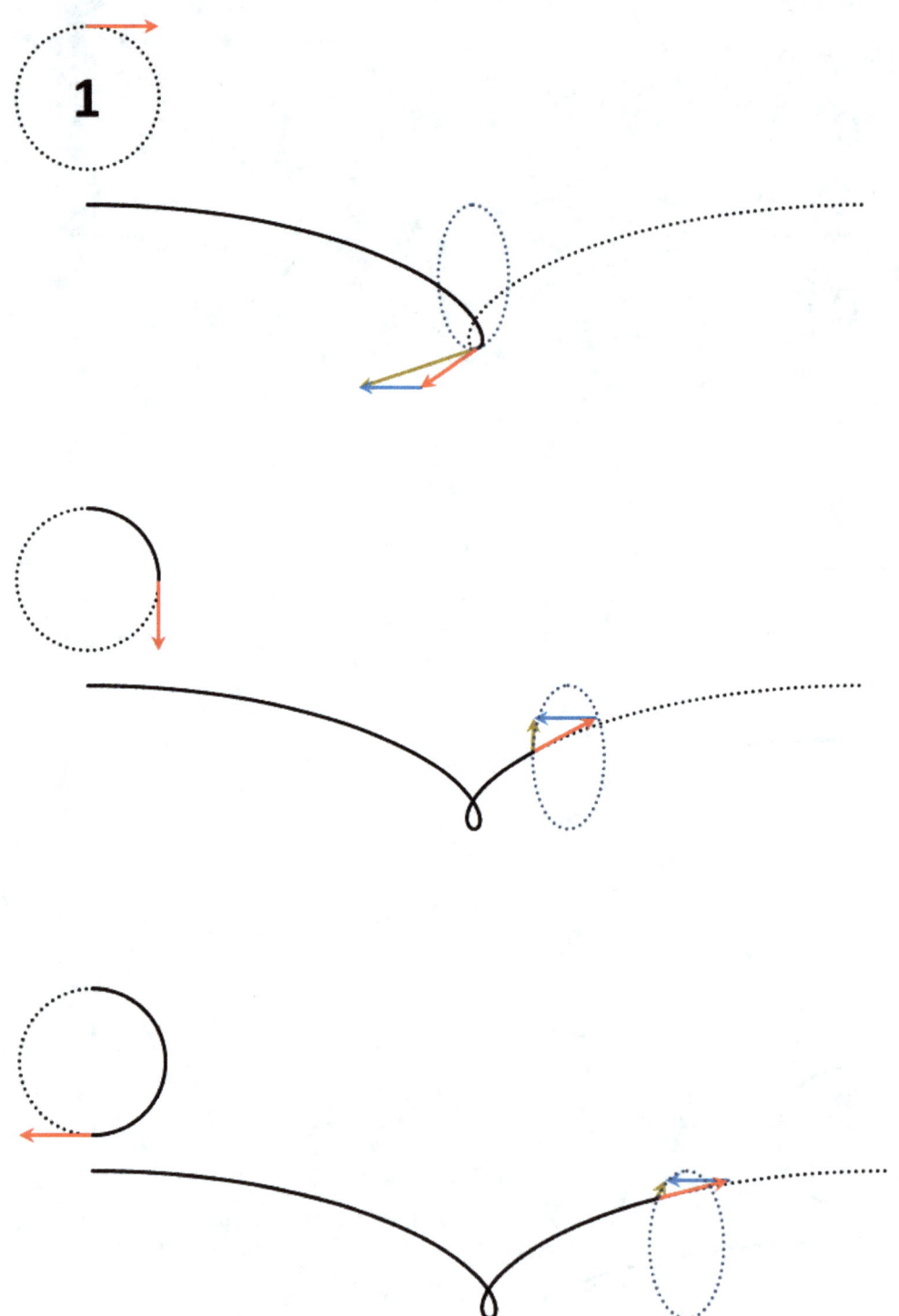

44

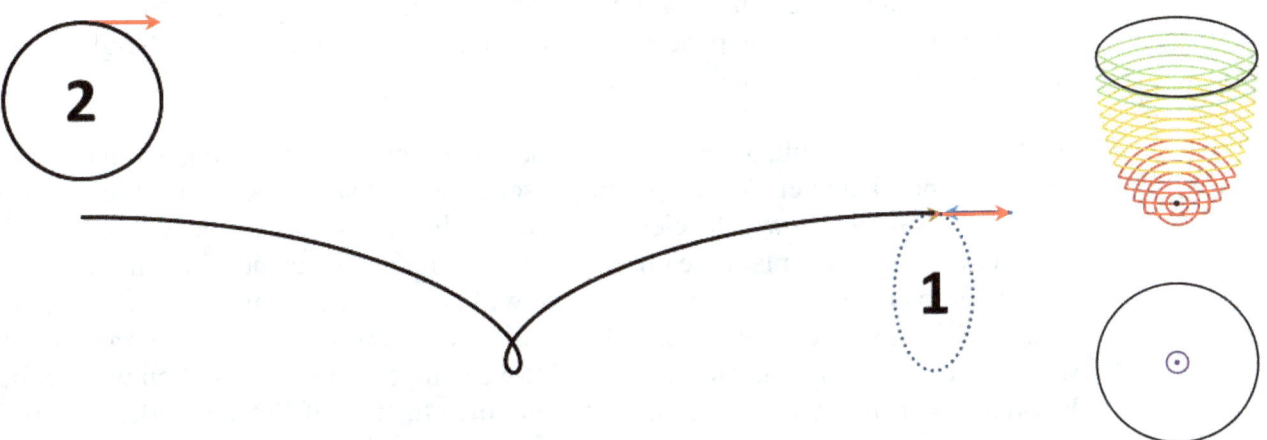

In this simulation, we compare a bug moving around a circular pattern that is *not* moving forward through space (upper circle in each picture) with a bug which is traveling at the same speed through space around an elliptical pattern which *is* moving through space. Again, here, we are looking at the case of velocity =.866c, in which gamma=2. In other words, the time slow-down is by a factor of two, and the ellipse height ratio is also 2. We can see that the bug moving on the non-moving circle makes two trips around its circle while the bug flying around the moving ellipse makes just one trip around the ellipse. Since the bugs both travel with the same constant velocity c through their local space, the length of the path of the bug tracing the moving ellipse as observed from the ground is exactly the same length as two trips around the stationary circle.

I'll add one more observation regarding this particular pattern. If you watch this "movie" it seems quite loopy: the bug spends much more time on the upper half of the pattern (going forward) and it passes the bottom point (going rearward) very quickly. However, I added a feature (not shown here) in which the bug sends an "I'm here" signal at various positions around the ellipse. If these signals are sent at equal angular positions, say every 30 degrees, and if these signals travel as light would, following the same rule of always moving at c through space (air in the metaphor,) then these signals would reach the pattern's central (antenna) point at equal time intervals. So, if one were moving along with the pattern, sitting at the antenna, and "looking" for these signals to trace the bug's movement around you, the bug would "appear" (be seen and measured) to be moving with a uniform (not loopy) motion around you.

From examining these various possible types of cyclic motion with our bug metaphor constructive models, I hope I have convinced you that ANY cyclic motion of energy that is trapped in a pattern which is moving through space would produce the time slow-down by a ratio of gamma. This slow-down in the rate of cycling is directly produced by the fact that the energy has a fixed speed through space, and part of this speed must be dedicated to tracking with the motion of the pattern, which reduces the cycle time. For

the cyclic pattern to be maintained, we also see that any component of forward and rearward motion in the pattern which would be circular if the pattern were not in motion must be elliptical when the pattern is in motion through space, with the height to depth ratio of the ellipse also equal to gamma.

We have started to build a constructive model that underlies the phenomena of relativity, but we are not there yet. We are getting a sense now of what may cause the time dilation and length contraction which scientists measure, but we do not yet have a model which is complete enough to explain the observed measurements of science. We have one major hurdle yet to come. After we master it, we will be able to explain why scientists always measure light to have the same speed regardless of the motion of the reference frame which is used to make the measurement. If we can get to that point, then we will be at Einstein's starting point. If we can get there, then the rest of Einstein's derivations of the equations of Special Relativity will follow. Only this time, we will have a constructive model which underpins it. We will have a vision in our mind's eye of a physically logical process that supplements the abstract mathematical description.

CHAPTER 4 – SYNCHRONICITY AND TIME SKEWING

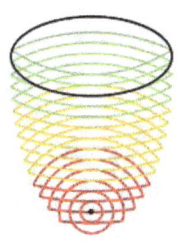

Like most amazing things
It's easy to miss
And easy to mistake
But when things are really great
It just means everything's in its place.

- Aimee Mann - I've Had It

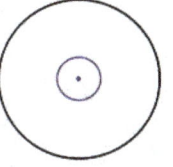

To get at this third important piece of our relativity constructive model, we need to go back into the realm of philosophy for a while. Einstein made a philosophical choice that was based on his values as a scientist which resulted in the way he described relativity, and ultimately, the way all scientists since have described reality. Because scientists value measurements, they made a practical choice which has affected their interpretation of relativity and the way they explain it when they teach it to others.

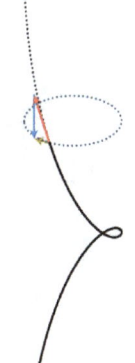

You may have noticed in my earlier discussions that I have assumed that space and energy simply exist, just as bugs and air simply exist. In my proposing the bug swarm model as a metaphor for trapped energy as particles of matter, I also simply assumed that time existed without further explanation. I also said that each particle had its own rate of time and suggested that any living beings living in a state of motion would not be able to tell that their own personal rate of time was slowed down relative to some other standard. Everything would seem normal to anyone enclosed in a spaceship traveling at a very high speed, and neither the time rate nor the length changes would be visible or detectable in any way within that environment. You might now be saying, really? If a basketball was squashed into a spheroid, how could I not see that? But if you carefully consider the fact that your eyeball and the lens within would also be flattened in the same proportion and the same direction, and your arms and fingers would also be shortened in the direction of motion, the basketball would look and feel perfectly round to you in your moving world.

It is this very fact that temporal and dimensional changes would be unperceivable to moving observers which led Einstein to abandon the notions of absolute space and absolute time. For him, since it was impossible to measure one's own time rate and length measurements against an absolute standard, he disregarded the concept of an absolute standard. The observer living on a planet in a solar system in a galaxy cannot escape from the reality that he must be moving, yet he cannot measure his rate of motion through space by any objective measurements of time and length in his moving reference frame. This was the lesson learned from the famous Michelson-Morley experiment.

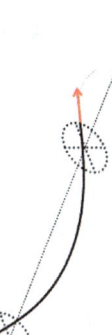

Michelson and Morley in 1887 attempted to measure the speed of the Earth through space with a very clever and sensitive experiment. Michelson and Morley at this time were naturally unaware of any length contraction or time dilation effects. They assumed that times and distances were absolute in the Newtonian sense. So, they planned their own sort of 'bug swarm' experiment on a macro scale. Beams of light were sent on round-trip paths of several meter's length. The light beams were split so that the two paths would be of equal length but at right angles to each other. The experiment is in fact quite analogous to the "four bug model" and the "bug swarm" in the horizontal plane.

As mentioned earlier in the bug examples, if the perimeter of turn-around of the bugs was a perfect circle, the bugs would not return at the same time if the pattern was moving through space. Michelson and Morley intended to measure the time difference in the return of the light beams, and in so doing deduce the speed of the pattern (the experimental apparatus, and with it, Earth) through space. What they did not understand, however, was that on the micro scale, the experimental apparatus, being composed of particles and atomic structures (which were trapped forms of energy traveling at the speed of light through space) would be assuming slightly non-spherical shapes. The effect of this was an unperceivable shortening of the apparatus in the direction of motion so that the two light beams in the macro scale experiment would always complete their round trips at the same time. This is important: this shortening was not unperceivable because it was very small; it was unperceivable because EVERY physical measuring device would shrink in the same direction by the same ratio. Their instrument was in fact sensitive enough to measure the tiny time differences that would result in this experiment due to the speed of the Earth around the Sun. But they implicitly assumed the apparatus was rigidly composed of hard little nuggets of stuff that would not deform in this way we've been describing due to the speed of the apparatus through space.

From this and many other experiments, it was repeatedly demonstrated that measurement of the speed of the Earth through space was not possible in a laboratory. But when Lorentz attempted to explain physically why such a measurement was not possible, his ideas were rejected. Lorentz said that the length of the apparatus *must be shrinking* by the factor gamma, but he had no reason why that should be so, other than it MUST be so for Michelson and Morley to get the null result that they did. Since Lorentz did not have the perspective which we now have that structures made of particles are composed of energy which also travels at the speed of light, Lorentz's explanations seemed contrived only to match the experimental results. He had no convincing physical reason why the measuring apparatus, assumed to be composed of solid nuggets of matter, should be deformed by this amount.

Now we are ready to start talking about this 'third' aspect of relativity which I mentioned. It does not have a popularly known name as do the other two aspects which are called 'time dilation' and 'length contraction.' I will call this phenomenon 'time skewing.' The Michelson Morley experiments showed that when light traveled in perpendicular round-trip paths over equal distances, the two light beams would return at the same time. This required some sort of physical "length contraction" in the direction of motion through space.

From our discussions earlier, it should also be clear that when the Earth was moving more rapidly through space, the time for this round trip would take longer than when the Earth was moving more slowly through space. Other experiments measured the round-trip speed of light under a variety of conditions, and also found the speed was unvarying.

However, since the clocks that measured this round-trip time were also moving through space with the measuring equipment, and the clocks were also constructed of sub-atomic particles which experienced the same slow-down from their motion through space at the same speed, the laboratory measurements of round-trip times for light always produced the same numerical result. The round-trip speed of light was always measured to be the same.

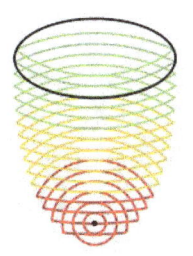

But Einstein took it one step further. He said that the ONE-WAY speed of light is always the same, regardless of the speed of the apparatus that measures it. That is quite a statement. Look again at the horizontal bug-swarm plots that show a harmonious return of all the bugs at the same time. It is quite clear that in those graphics with v > 0 that some bugs spent much more time moving in one direction than they did after they turned around. Isn't the bug that is moving to the right when the pattern is moving to the right moving at speed c-v relative the pattern's center? And isn't the bug moving to the left when the pattern is moving to the right moving at v+c relative to the pattern's center? Although the round-trip average speed is the same, wouldn't the one-way speeds as measured in the moving frame be quite different? How is this reconciled with Einstein's statement that the one-way speed of light is always measured to be the same?

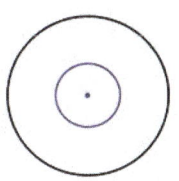

The answer is time-skewing. And this concept is central to all that will follow. For the moment, let's follow Einstein's logic, and consider the practical aspects of experimental measurement. Remember that Einstein's statement is that the one-way speed of light is always MEASURED to be the same. Remember also that measurement is the primary scientific value. Einstein says that if you want to measure the speed of something, you need to have two clocks in different places that are synchronized; that is, they must not only run at the same rate, they must also read the same time "at the same time". What does that mean? And how is that accomplished, in practical terms? There are several ways to do it. You might place two identical clocks side-by-side, set them to the same time, and then separate them by moving one clock to a different location. This method seems simple and straightforward, but in fact there are some very subtle aspects to it that need to be discussed in detail. We will come back to this method of synchronizing separated clocks later.

The second way which clocks that are separated in space can be synchronized is considered the standard Einstein clock synchronization procedure. It also seems rather practical and reasonable, but when viewed in light of our constructive model, I think you will see that this method also contains a boatload of irony. Einstein's method of clock synchronization is as follows: two clocks are tested to verify that they run at the same rate, then the clocks are spatially separated. One clock is at station A and the other at station B. To synchronize the clocks with Einstein's method, at a specific recorded time on clock A, station A sends a signal, a light signal or a radio signal, to the clock at station B. When station B receives the signal, it records the time on its clock, and it also immediately sends a return signal back to station A. Station A records the time when the return signal is received. Next, station A and B communicate with each other about the recorded times.

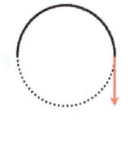

Clock A has recorded two times. Clock A and clock B are said to be synchronized by Einstein's method if the time recorded by clock B is exactly mid-way between the two times recorded by clock A. Now the irony should be apparent. Because a light signal or a radio signal (which also travels at the speed of light) is used to synchronize clocks in this way, if two clocks synchronized in this way are then used to measure the speed of light between A and B, the ONE-WAY speed of light will be automatically guaranteed to be the same numerical value in both directions.

The fact that the two-way speed of light is the same in all directions is a rather remarkable fact which occurs, as we have seen, because the structures of measuring instruments (measuring rods and clocks) are composed themselves of energy which must play by the same rules as the light being measured. The assertion by Einstein that the one-way speed of light is the same in all directions was a bold move indeed. We will come to see how he made this argument convincing, and we will see that what I'm calling time-skewing is exactly the correction required to make the one-way speed identical to the two-way speed.

Look again at the third bug swarm plot, in which v=.866c and Gamma = 2. Suppose that clock A is at the center of the pattern, and clock B is at the rightmost point. In other words, clocks A and B are moving through space very rapidly with clock B "leading" clock A, that is, clock B is "forward of" clock A spatially as they travel through space toward the right. In this plot, we see that the forward-moving bug leaves clock A at time zero and gets to clock B on the 28th clock tick (out of 30.) Then it returns to clock A on the 30th clock tick.

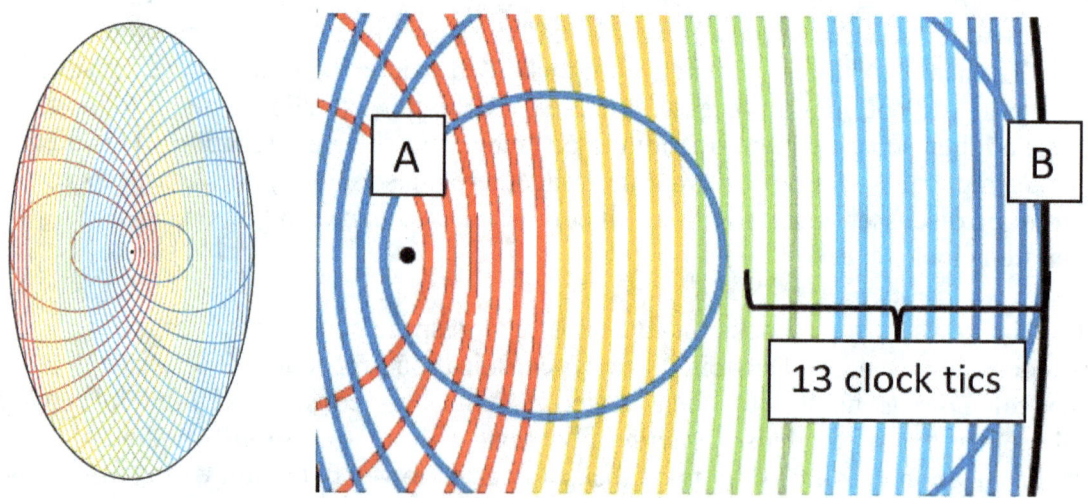

With Einstein's synchronization procedure, we expand the consideration of light moving through space to the macro scale. The clocks may be several meters, or several miles, or several light-minutes apart. Clock A sends its signal at noon exactly, and records noon+30 ticks when the return signal from B comes back to it, then tells clock B that his clock should have read noon+15 ticks when he received his light signal from A. If clock B for some reason recorded noon+28 ticks when it received the signal, station A tells B to set his clock back 13 ticks and try again. At 1 pm exactly, A sends another test signal. B now receives the signal at 1pm+15 ticks and sends a return signal which A receives at 1pm+30 ticks. Perfect. The clocks are now synchronized per Einstein. At two o'clock they are ready to measure the speed of light with their synchronized clocks. Guess what? The speed of light so measured is the same in both directions, numerically equal to the distance between them divided by 15 clock ticks.

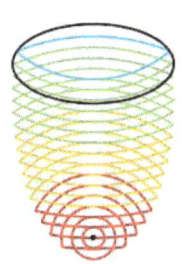

Let's make this thought experiment more concrete. Suppose a rocket ship happened to be at, or nearly at v = 0 in its local space, and it precisely synchronized identical clocks at various places (Clock A at midship, clock B at front of ship.) In this v = 0 state, clock B would receive a signal from A at tick 15 on its clock (synchronized) – In the metaphor, this represents reception of the mid-green 15th tick in our v=0 circular bug plot.

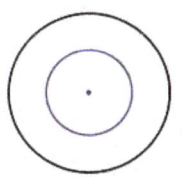

If this ship then left its clocks untouched and accelerated up to a very high (v/c = .866) speed through space, and then went back into "coast mode," it would find itself metaphorically in the situation of the v/c=.866 (gamma = 2) bug plot shown above. All the clocks would be running slower, but now clock A would find it was receiving A's test signal at 28 tics after the expected test hour, instead of 15 tics after the hour. When B immediately sent the return signal, A would receive it on its 30th tic, but as they discussed these results after the fact, clock A's manager would later tell the manager of clock B that to be synchronized, it should set its clock back 13 tics. Then on the next test, B would get the test signal at 15 ticks after the test hour, and the clocks would again be synchronized.

You can count on the v/c = .866 (gamma = 2) bug swarm plot that there are (obviously) 13 tics between the green 15th ring and the dark blue 28th ring shown reaching the front of the ship. Mathematically, the precise v/c = .866 which yields an exact gamma of 2.00 is $\frac{\sqrt{3}}{2} \simeq .86602$ But notice that this is very close to 13/15 $\simeq$ .86667. This is not a random coincidence. The amount of time skewing clock offset is precisely v/c, which is approximately the (28-15 =) 13th tic out of 15 tics in our bug swarm clock. The amount of time-skewing is equal to the v/c velocity ratio. If the space ship were to move to an extremely high velocity such as to v/c = .9999, the time skewing would be 14.9985 out of 15 tics. In other words, it would take light 29.9985 tics to get from A to B, and only .0015 tics to go from B to A. It could never take more than 30 tics for the signal to go from A to B because its round trip will always be 30 tics! As we know from our bug metaphor, v can never exceed c!

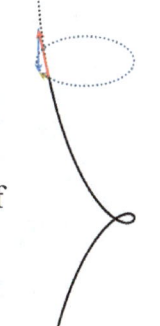

A similar thing happens if clock A synchronizes with another clock C which is at the left in our diagram. Clock C is trailing behind clock A in their mutual motion through space. But now, the time skewing runs the other way. In terms of what we see in our diagram, clock C, the trailing clock must set its clock forward 13 ticks in absolute terms in order to be synchronized with clock A. And now, if you think about it, you will see that clocks B and C are also synchronized, too. If they send signals to each other, they will find that no matter who sends the signal first, the other clock will read a time which is midway between the two times of the original signal sender. In fact, this will hold true between any two Einstein-synchronized clocks at any positions that are travelling along at fixed positions in the pattern, whether in the horizontal plane or not. What Einstein has established is a practical, consistent method of synchronizing clocks in a moving

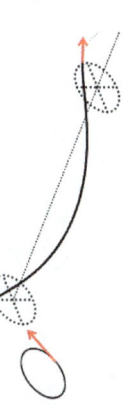

reference frame. And with this method, the speed of light will be numerically the same when measured between any two clocks so synchronized that maintain stationary positions in that moving frame.

If you're like me, you might be asking, why did Einstein select this method? It seems bogus on the face of it, like a self-fulfilling prophecy, or more kindly perhaps, a truth by definition. Sure, the speed of light will be measured to be the same in all directions if you set clocks forward and back just to make it so. Well, Einstein was a practical guy, and in fact, his method is the only practical way to do it. If you could somehow know your absolute speed through space, or if you could communicate instantaneously, without having to wait for a signal to exchange information, then you could set your clocks "properly" so that the speed of light would be measured to be different in different directions relative to your reference frame, as we see it is in our swarm plots. But in fact, the Michelson Morley experiments and others showed that one cannot measure one's own absolute speed through space. Since we have no means for "instantaneous" communication, the Einstein synchronization procedure is the only practical method available.

Ah, but wait, you might ask, what about the other method of synchronization that you mentioned, and skipped past? Why not simply synchronize two clocks side by side, and then move them to locations A and B? Wouldn't that method allow an experimenter to detect the different travel times for light in the forward and backward directions? To this, the answer is no, and again I ask your patience, because I will return to discuss this later. For now, just trust me that this method will not work either.

By now, I'm sure you've figured out that what I mean by time skewing is this notion that in the realm of practical measurements within moving reference frames, the forward clocks are set back in time, and the trailing clocks are set forward in time. The standard against which they are 'set forward' or 'set back' is the <u>notion</u> of absolute time. To Einstein, and to Physics, they are not set forward or set back: they are simply set to be synchronized in the observer's frame of reference. This notion of 'absolute time' is one that physicists recoil against, and I know that. It is, however a useful idea in the discussion of how relativity works physically.

It may have value at this point to pause just for a moment to inject a concept that may be helpful. I call this concept State Universe. The physicists seem to have this in mind when they talk about the Universe shortly after the Big Bang. They describe the State of the Universe several nanoseconds after the Big Bang and the State of the Universe several hundred thousand years after the Big Bang. If the Universe is continually evolving scenario of energetic transformations, doesn't it follow that there is still a "State of the Universe?" It seems obvious that the Universe operates in parallel in the general sense; that is, different things must be going on in different places 'at the same time.' Mars orbits the Sun *while* Earth orbits the Sun; Mars does not wait for Earth to move before it takes its turn moving. As a bird lands on a branch in Mexico, a child turns the page of a book in Italy.

But because relativity correctly shows that different observers may observe events, and even the time sequence of events differently, Physics seems to deny that there is any reality to the concept that there is an 'actual' sequence of events corresponding to events that occur relatively close together in time but relatively far apart in is space. But if there is a State of the Universe in which event A has occurred, and event B has not yet occurred, I argue that A actually (absolutely) occurred before B in "State Universe." This does not deny that an observer may correctly record B as happening first in their frame of reference. Nor can it be proven who is correct when two observers make valid measurements that conflict regarding whether A or B occurred first.

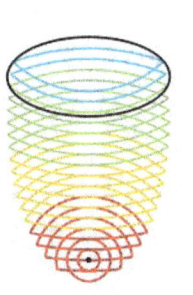

It is this notion of State Universe that I am referring to when I suggest that a forward clock and a rearward clock in a moving reference frame do not read the same time *at the same time*. It is this same notion that I use when I look at the bug swarm plot and say that the rearward moving bug turns around in its cycle before the forward moving bug does. This is the key to time skewing in Special Relativity: because of Einstein clock synchronization in a moving frame, to the observer in the frame moving through space the bugs turn around (the light signals turn around) at exactly the same time (in their own frame of reference!) But to an observer not sharing the same frame of reference, they generally do not turn around at the same time.

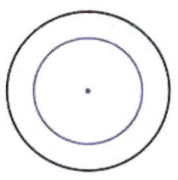

To demonstrate this time skewing further, I modified one of my bug plot simulations to show clock times at the forward, central and rearward locations of a moving cyclic pattern. Shown below are several stills taken from this simulation. In these plots, the numbers shown represent clock times at various positions. Rather than count 30 ticks for a complete cycle, and 15 tics for each one-way part of the cycle, such that the time skewing is (.866) *30 = 13 tics, here the complete cycle time will be 2 clock ticks, the one-way time will be 1, and so the time skewing is (.866) *1 = .866, which is rounded to .87 in the graphics.

Notice specifically that by skewing the clock times in this way, the 'arrival times' at the left and right positions on the ellipse are equal to "1" (highlighted in red) in both cases. This is the requirement for Einstein clock synchronization. The signal sent from the center at time "0" must arrive at both the forward clock and at the rearward clock at time 1. In order for this to happen, the clocks must be offset in time from the central clock. Notice that the 'leading' clock must initially be set back to -.87, and the 'trailing' clock be set forward to +.87 when the central clock is at 0. As the ellipse pattern moves, the time-skew of .87 is maintained as all the clocks in the moving advance in time at the same rate. But notice that this rate of time advance is at half the rate of the time advance in the stationary circle. This is most easily seen by noticing that the central clock of the moving ellipse runs at half the rate of the stationary clock. This happens because in both systems, the central clock must read "2" at the end of one cycle. We see that the pattern moving at .866c takes twice as long to complete its cycle than the stationary pattern, because at v/c = .866, gamma = 2.

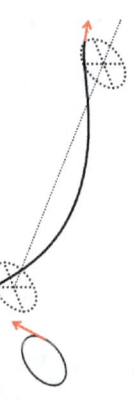

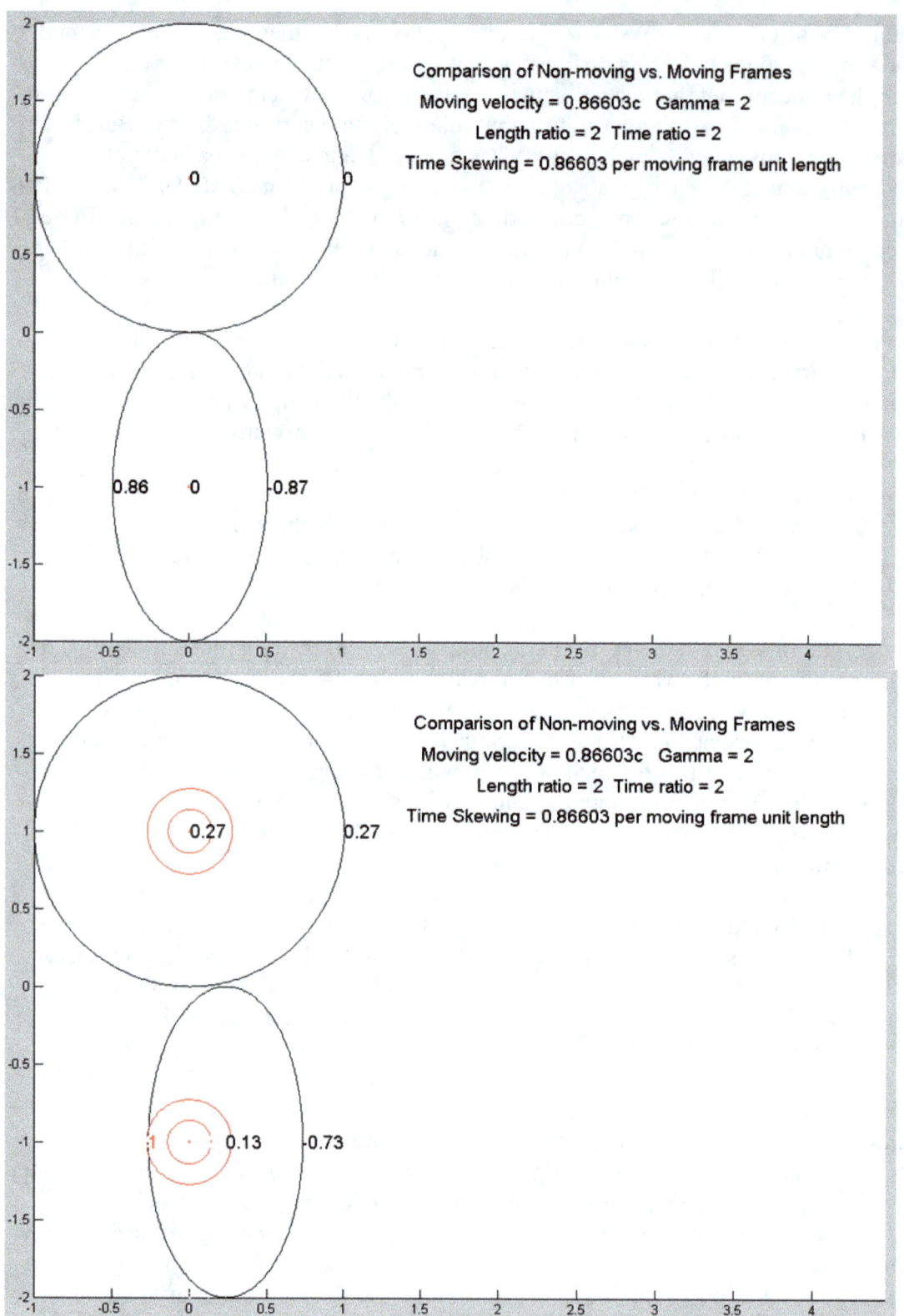

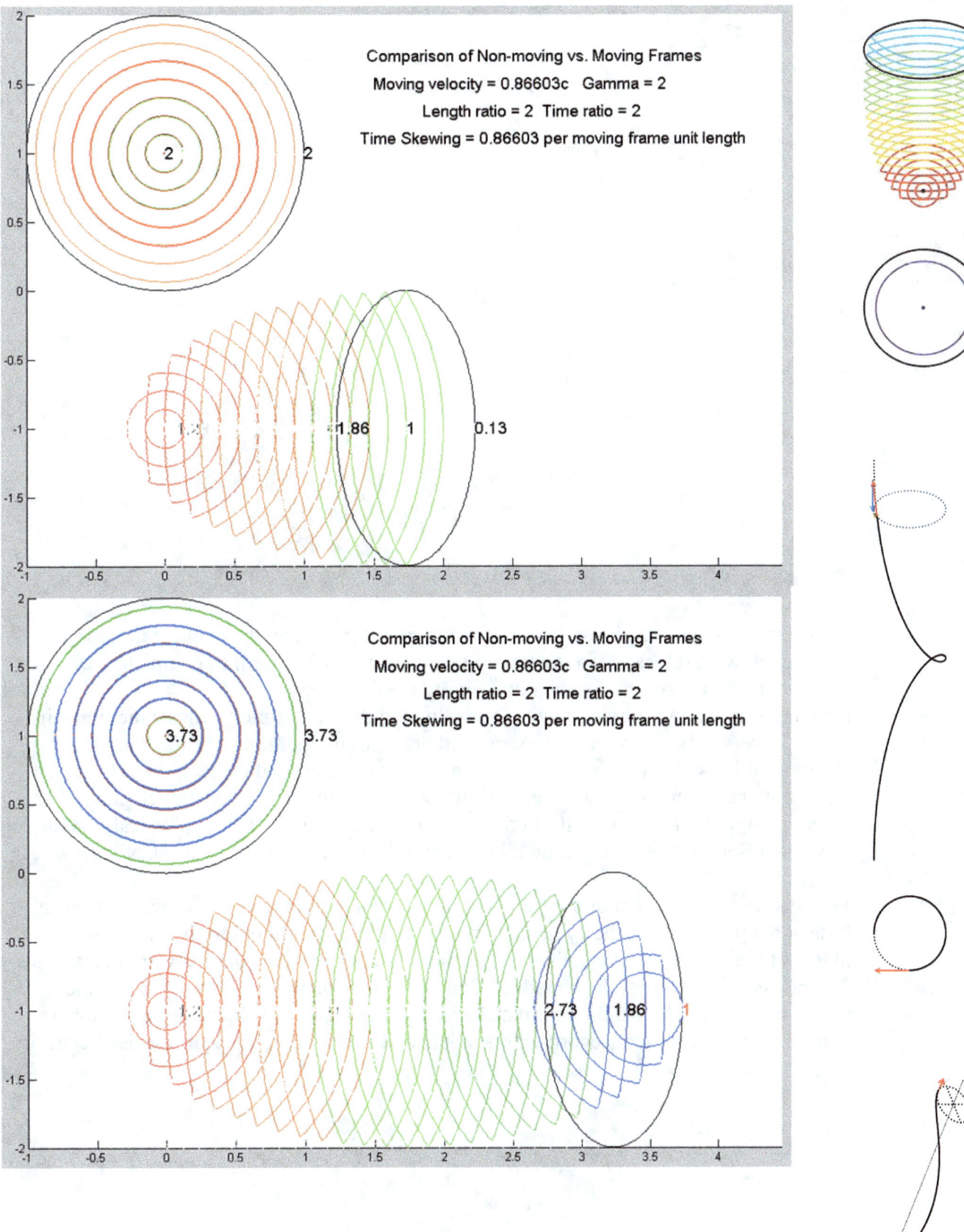

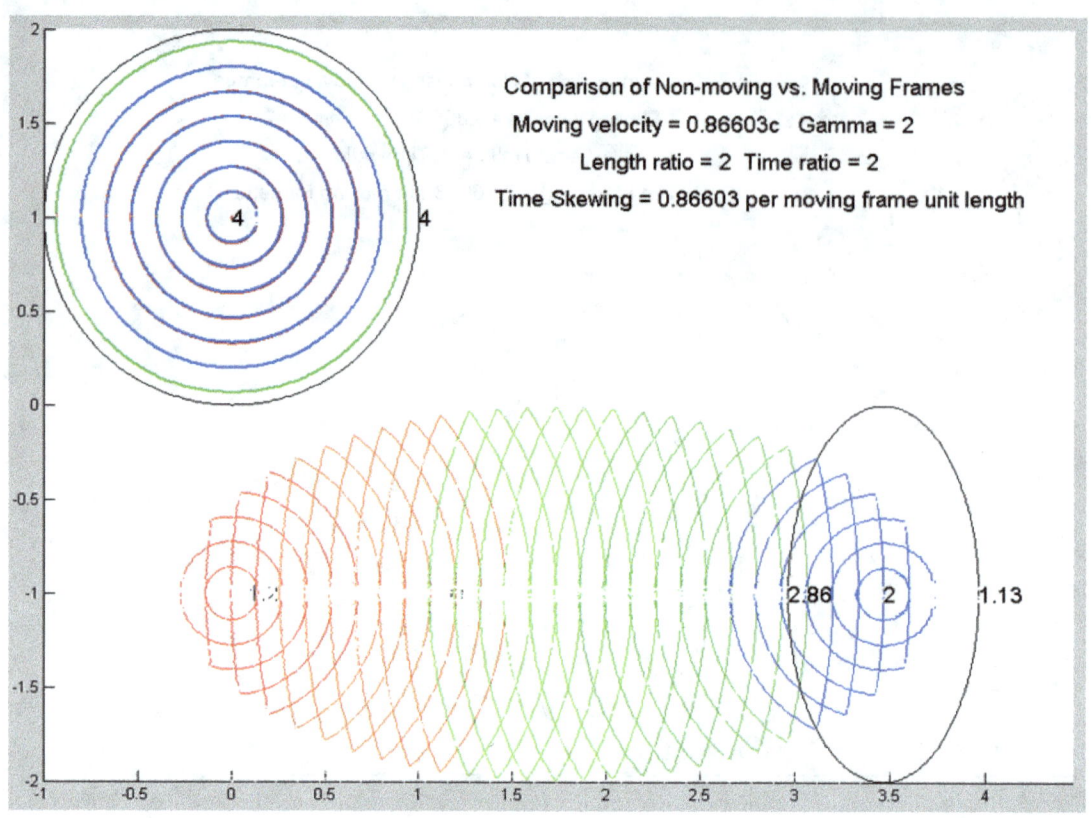

Notice in the sequence above that for the non-moving pattern, the cyclic pattern is completed twice (clock=4) in the time it takes the moving pattern to complete 1 cycle (clock=2). Note also that for the non-moving pattern, there is no time skewing, that is, all the clocks in the non-moving pattern actually read the same time "at the same time" in State Universe. But in both the moving and non-moving patterns, we see that the criterion of Einstein clock synchronization is met: the rearward clock and the forward clock both read time = 1 when the light first gets to them, and the central clock reads 2 when the light first returns to it. In this sequence, we see represented physically all three aspects of Special Relativity: time dilation, length contraction, and time skewing.

I pointed out that in the moving frame, the forward and rearward clocks were offset in time at the beginning of the cycle, and by an amount proportional to the ratio v/c = +/- .866. At the end of the cycle, the central moving clock reads 2, the rearward clock reads 2+.866, and the forward clock reads 2-.866. The time skewing is always maintained, as all the moving clocks wind forward at 1/gamma the rate of the clocks in the stationary pattern. This is in perfect accord with Lorentz and Einstein's theory of Special Relativity.

CHAPTER 5 - THE MEANING OF NOW

There will be more discussion and constructive-model-based demonstrations of Special Relativity ahead, but for now I want to touch on the statements I made earlier about truth. In the last chapter, I discussed the notion of "State Universe" regarding the possibility of different observers making different assessments of the sequences of events they observe because of time skewing associated with their motion through space.

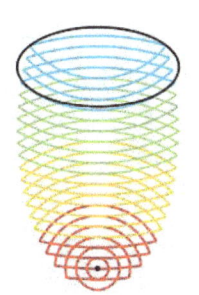

Physics seems content with a philosophy that says when such conflicts arise, **both** observers are correct. Because science places its highest truth value on measurement, if both measurements are correctly performed, both observers are considered correct although the observers may disagree in their conclusions. Truth itself is considered relative to the observer, and the existence of an absolute truth is entirely denied. But a more harmonious philosophy might say that there IS an absolute truth about which neither observer can know.

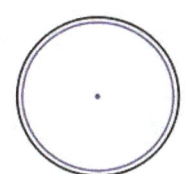

I watched a popular series that discussed relativity on television recently. Something that was said in it was quite remarkable, and I think that now is a good time to discuss it. With our constructive model, if each observer in the universe has his own personal frame of reference which is developed according to Einstein's synchronization procedure, then each observer will naturally co-ordinate their own frame of reference by setting remote clocks "forward" and "backward" (in the State Universe sense) at various locations such that the speed of light is measured to be the same in all directions, as we have just discussed. But if this notion of State Universe is rejected in favor of the notion of multiple relative truths, this can lead to some remarkable and I think ridiculous statements about reality. It was said on this television program that someone very, very, far from Earth, if traveling on a bicycle toward Earth, would have a time-skewed personal frame of reference such that when they said "NOW", that their NOW would be simultaneous with a time in our past. True enough. Even at their slow bicycle speed, there would be some very tiny time-skewing which would, over billions of light years of distance amount to a significant time-skewing. If they were far enough away, their NOW could be simultaneous with our American Revolution in 1776. If, however, the bicycler turned around to travel away from Earth, their NOW would coincide with a time in our future, say the year 2265. This is conceptually compatible with what we have discussed, because forward clocks are set backward in time, while trailing clocks in a reference frame are set forward in time, in the State Universe sense.

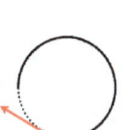

The speaker then discussed a concept of a 'NOW slice,' an idea in which each observer has a sense of 'the present' which cuts across the universe of space and time, or as they said, each observer slices space-time in a different direction depending on their velocity through space. While I'm sure this is mathematically correct and consistent with relativity as physics expresses it, I think it perfectly suits my argument about the nature of reality, and the nature of truth. The television program's physicist claimed that relativity therefore showed that the past and present are not separable, and that in a way, our future already exists, because someone far away conceives themselves as simultaneous with it. That is absurd.

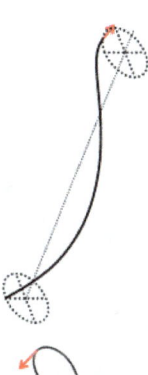

57

What came to my mind was a scenario in which two bicyclers in that very remote galaxy approached each other, one moving away from us and the other moving toward us, and as they passed each other, they both said 'NOW.' Their moment of passing would truly represent an event in the universe, but what moment on Earth would be simultaneous with that remote event? I think the answer to that question puts everything I'm trying to say in sharp focus. In the absolute sense, that event of their passing would be truly simultaneous to only one State of the Universe, and therefore only one state of the Earth. Both bicyclists cannot be "right" in their version of simultaneity with events on Earth, since the Earth cannot be in two very different states at the same time. Both bicyclists can claim their own relative truths based on their primal valuation of measurement by synchronized clocks in their own reference frames, but rather than say that both of their relative truths are "right," I prefer to say that there is one absolute truth, and neither of them can know it. I think it makes more sense to say that, in the unfolding scenario Universe, there is only one state of the Earth at the moment the two remote bicyclists pass each other. Conceptually, if the cyclists hold their own measurements as their highest truth, then their truths are relative, different, and in conflict. Science almost seems to believe that Earth might be considered to be in two different states at that moment of cyclists passing. But we live on Earth, and we are quite certain that events of 1776 and 2265 are not, and never will be occurring here simultaneously.

Science, through relativity, has been saying that truth lies in measurement, and since equally valid measurements may draw different conclusions about the history of the Universe, truth is relative - there is no absolute truth. I'm asking you to consider the possibility that there is an absolute truth, an absolute history of the Universe and what has happened in it, but it is a fundamental characteristic of the Universe that no scientific measurements can determine precisely what that history is.

CHAPTER 6 - SEPARATING SYNCHRONIZED CLOCKS

Now if you've been thinking and remembering where this discussion of time skewing started, you might ask, what about that other method of clock synchronization you mentioned, but skipped past? Why can't we just synchronize two clocks side by side, move them apart, and then measure the speed of light in different directions? If the clocks are synchronized that way, then surely, they will be able to detect that it takes more time for light to get from A to B than from B to A? Well, actually… the answer to that is no, and the reason for this is subtle but also very important. While at first glance this time-skewing Einstein clock synchronization procedure seems like a handy self-fulfilling prophecy, there is actually something to it, physically. Something quite profound.

Recall for a moment that we said that when traveling through space at speed v, the time taken to complete a repeating cycle would slow down, whether a micro energy pattern cycle or a macro pattern light beam. Understand, however that this slow-down would be relative to the cycle time if not moving through space, or we could say, relative to an absolute standard of time in the universe. But immediately after that, we said that this slow-down of cycle times could not be detected by an observer and their measuring devices that were traveling along with the pattern. For light moving in a transverse direction within the pattern, the distance would be unchanged, but the speed of light would be measured to be numerically the same because the clocks measuring the time would be slowed down in the same proportion. And for light traveling on a round trip parallel to the direction of motion of the pattern, the length of the trip would be shortened (the length measured with a meter stick which was similarly shortened) so that the cycle time was coincident with the lateral round trip, and the clock again would be slowed down, making the measurement of c numerically unchanged.

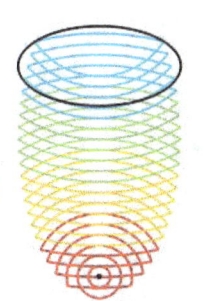

When measuring the speed of light in the direction of motion on a one-way basis, we have all three aspects of relativity in play. The measured distance is shortened by the flattening of the atomic structures involved, the clocks slow down, and the time-skewing of Einstein's synchronization makes the forward and rearward times come out again numerically identical. That is fantastic. Now, why can't we overcome Einstein's sneaky self-fulfilling prophecy of setting the clocks forward and back which seems to rub out the difference between forward and rearward times? The answer is that, just as the clocks cannot escape the time-slowing involved when they stay in motion with the pattern, the clocks cannot escape the time-skewing if we synchronize them and then separate them. This is an amazing fact, and it took me some time thinking on my own to convince myself that this is true and figure out how this works.

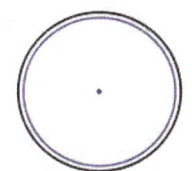

What I've just suggested is that clocks that are synchronized side-by side and then separated will remain synchronized, but not in the absolute sense that would make measurement of two different one-way speeds of light possible. To the contrary, when side-by-side synchronized clocks are gradually separated from each other to different positions within a reference frame, these clocks will remain synchronized <u>in the Einstein sense</u>. In effect, they will set themselves back in time or forward in time such that the Einstein synchronization rule is maintained! And if these clocks are then set in fixed positions in the reference frame and used to measure the speed of light between two points in the reference frame, the measurement of the speed of light will STILL be numerically the same as always, regardless of the direction of the measurement!

I have to be a little careful here, because physics is a very precise business. You perhaps noticed that in the last paragraph I used the word *gradually* in describing the separation of the synchronized clocks. This was entirely intentional. If one wants to get very picky about it, the truth is that the described Einstein synchronization will be maintained only approximately, but the precision of the synchronization can be made as near to perfect as one wishes by moving the clocks as slowly as one wishes. Or, as the mathematicians would say, "in the limit as the separation velocity approaches zero", Einstein synchronization will be perfectly maintained.

Absolutely perfect Einstein synchronization can only be achieved if the clocks are separated at zero relative speed, which is impossible. That being said, the automatic time skewing of separated clocks is very real, and, considering the speed of light is extremely high by human standards, it is extremely easy to separate clocks slowly enough such that the precision of Einstein synchronization is perfect for all practical purposes (such as for

measuring the speed of light with them.) Now, how can synchronized clocks automatically adjust themselves in this dramatic way?

There are many types of clocks, of course, and explaining how this 'automatic time skewing' would occur in a mechanical clock would be, I imagine, very tricky. It would also be difficult to get into the details of how a mechanical clock would slow down when moving at a very high rate of speed through space. There are springs, and forces, and rotations which would make such an analysis difficult. But when we say that in very fast travel through space ALL processes are slowed at the physical level, including atomic half-lives, and chemical reaction rates, and synapse firings, we see no reason to believe that electric motors, automobile engines or any other mechanical device would be excluded from this reality. And so, I am satisfied here to demonstrate (with a thought experiment) how synchronized Light Clocks would be automatically time-skewed by separation in a frame of reference moving at a high rate of speed through space. A Light Clock keeps time by bouncing light between opposing mirrors. Since, as we know, the speed of light is always measured to be the same for any round trip, the clock keeps a steady rate of time, and each round trip "tic" takes a time equal to the distance between the mirrors divided by the speed of light. Put it in a vacuum, and you have a reliable clock. Very simple.

And here, I've considered two configurations of Light Clocks, one oriented longitudinally (in the direction of travel through space) and one oriented transversely (perpendicular to the direction of travel). In both these cases, the process that would maintain the clocks in Einstein synchronization is imaginable. The longitudinal arrangement is the simpler case, and I will not get analytic with calculations, but describe the ideas qualitatively. With additional thought, the reader can prove the numerical correctness.

Now with this longitudinal clock (aligned with the direction of motion), let's imagine two scenarios. First, suppose the clock is in a spaceship that is not moving through space. If this happened to be true, as we know, the light would actually take as much time to travel forward (say, from mirror R for 'Rear' to mirror F for 'Front") as it would take to travel rearward from F to R. So, if we slowly and steadily moved the clock as a unit forward within the spaceship, every time we moved it a little, we would be adding a little distance, and therefore time to its R to F leg… but equally removing the same amount of distance and the same amount of time from its F to R leg of its cycle. Remember this is true because in $v=0$ space, light *actually* travels the same speed in both directions.

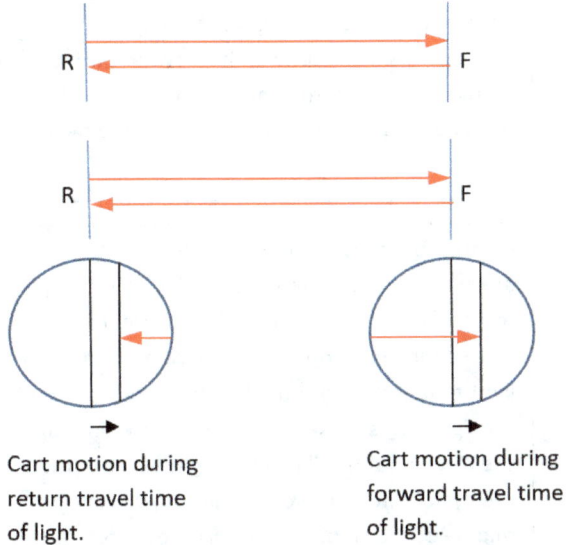

Cart motion during return travel time of light.

Cart motion during forward travel time of light.

When a ship is stationary (not moving through space,) light spends equal time in each half of its clock cycle.

On this stationary ship, two identical clocks are synchronized, then one clock is rolled forward on a cart. On each light cycle, a slow-moving cart moves the same distance during both halves of each light cycle. The extra time added to the forward light path is equal to the time subtracted from the rearward light path. The clocks on this non-moving ship therefore remain synchronized both in the Einstein sense, and in the absolute State Universe sense.

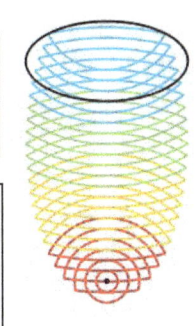
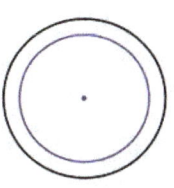

But now imagine scenario 2, in which the ship is moving forward very, very fast through space. We know from our bug analogy that the actual travel times in the forward and reverse direction will be very different. At our familiar Gamma = 2 speed of .866c, it takes light about 14 times as long to go from R to F as it takes it to go from F to R. When we move the clock slowly and gradually forward in this situation, we add much more time to the R to F leg than we subtract from the F to R leg, because light spends much more of its time moving from R to F. We are subtracting a little time from the rapid leg and adding much more time to the leg that takes longer. The clock therefore ticks a little slower when it is being moved forward in the ship. As a result, the clock that is moved forward is set back in time in an imagined "State Universe" sense relative to the Light Clock in the middle of the ship that was not moved forward. But calculations would show that the two light clocks would still be synchronized in the Einstein sense. This means that if the ship maintained its constant speed, and you then used these two clocks to measure the speed of light from R to F (or from F to R) in the ship, you would measure the same speed of light in both directions, despite the fact in the State Universe sense, the travel times would be unequal.

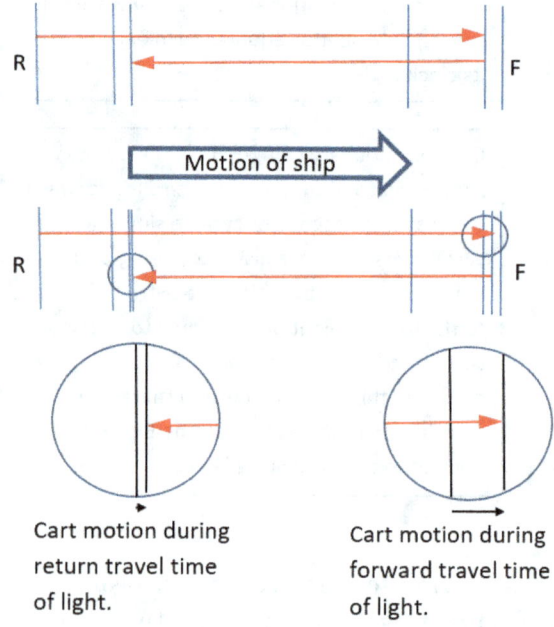

When the ship is moving fast through space, light spends more time on its forward path than on its rearward path. The round-trip cycle time is slower than the stationary ship by the factor gamma.

On this fast-moving ship, two identical clocks are synchronized, then one clock is rolled forward on a cart. On each light cycle, the cart moves further during the forward motion of the light than it does during the rearward motion of the light. This makes the overall cycle time of the clock on the rolling cart slightly longer than the cycle time of the clock that sits still in the middle of the ship. As the cart moves forward, the clock on it tics slightly slower, which sets the forward clock backward in time relative to the clock that stays at midship. This maintains synchronization in the Einstein sense, but not in the absolute State Universe sense.

Next, we may consider the case of two clocks which consist of light beams which cycle in the lateral direction, that is, a light beam which bounces in a direction which is perpendicular to the bulk motion of our ship through space. Again, we have two identical clocks synchronized right next to each other in the middle of the ship, and we separate them by moving one clock toward the front of the ship.

First again, assume that our ship is absolutely still (v=0) in space. The light in each clock is now traveling along one thin line, not only in not only in the reference frame of the ship, but also in space itself. Since the speed of light is about 186,000 miles per second, the light is bouncing from mirror to mirror, from port to starboard and back across our spaceship billions of times per second. Now, any motion of either clock will theoretically slow down its tick rate, but we would have to move it very, very quickly to register any detectable slow-down, even for these high-precision clocks. The clocks would tick slower if the ship was at high speed because the light would have to travel in a zig-zag path through space, and as we have discussed, the slow-down ratio is the ratio of the diagonal zig-zag distance to the straight across distance.

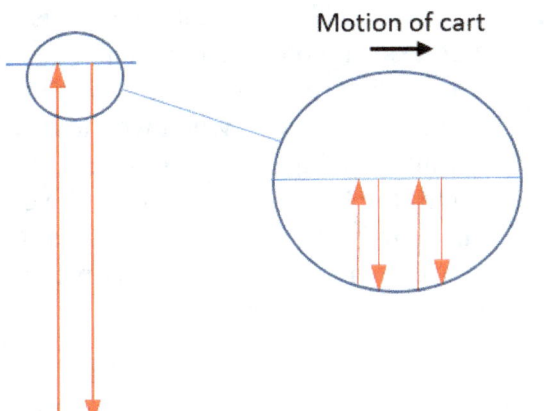
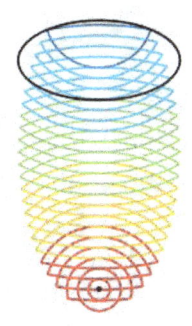
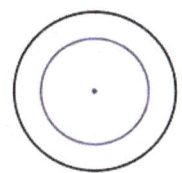

Transverse Light Clock in a non-moving ship. The motion of the cart is so slow compared to the speed of light that the angle of the light beams remain virtually vertical. The clock on the cart remains synchronized with the midships clock both in the Einstein sense and in the absolute State Universe sense.

But if the ship is still in space (v=0), and we move a clock forward at ordinary low speeds, rolling it on a cart perhaps, for all practical purposes, since the light is bouncing back and forth billions of times per second, the light is still going virtually straight back and forth across the ship. The distance that the cart moves forward in a billionth of a second is not enough to create anything but the tiniest sliver of a triangle, and for such triangles, the difference between the hypotenuse and the long side of the triangle is entirely negligible. Put another way, the sine of 89.99999999 degrees is .9999999999999999985. For all practical purposes, the clock on the cart ticks at the same rate as the stationary clock, and so these two clocks stay synchronized in the ordinary sense. There is no set-back or set forward of the times on these clocks. So, this is good. When v=0, and Gamma=1, atomic structure is spherical, lengths are not shortened, clocks are not slowed, and clocks synchronized side-by-side remain synchronized after slow separation, both in an absolute sense, and according to Einstein's rule. If we use these separated clocks to measure the speed of light, we will get the same result in both directions. No surprises here. This pattern is the first bug swarm pattern, where the bugs really travel outward and inward in true circles.

Now, let's try the same thought process assuming that the ship is moving really fast through space. Now our light beam clocks are running slower compared to their maximum possible rate because the beams are moving through space on a zig-zag path.

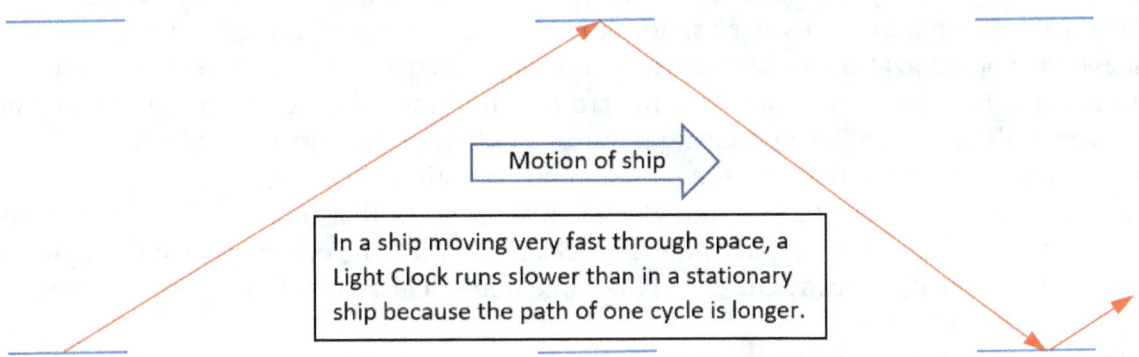

In a ship moving very fast through space, a Light Clock runs slower than in a stationary ship because the path of one cycle is longer.

If we are moving at v=.866c, Gamma=2, our clock is slowed, running at half speed in the State Universe sense. Note that at this speed, the angle of the zig-zag is 60 degrees, as shown above. Everything seems entirely normal and unchanged to us, though, because we are inside the ship, too, and our internal particle clocks, and our chemical and biological clock are all slowed down by exactly the same rate. So, what happens in this case when we separate two side-by-side synchronized clocks? Now, when we move a clock forward, even very slowly on its rolling cart within our ship, we are adding extremely minute slivers of angles to the zig-zag path of the light of the clock on the cart. But now, every miniscule move forward increases the length of the zig-zag path in a way which did not happen when the ship was at rest in space. No matter how slowly we move the cart, we have a proportionally tiny but appreciable increase in the length of the zig-zag path.

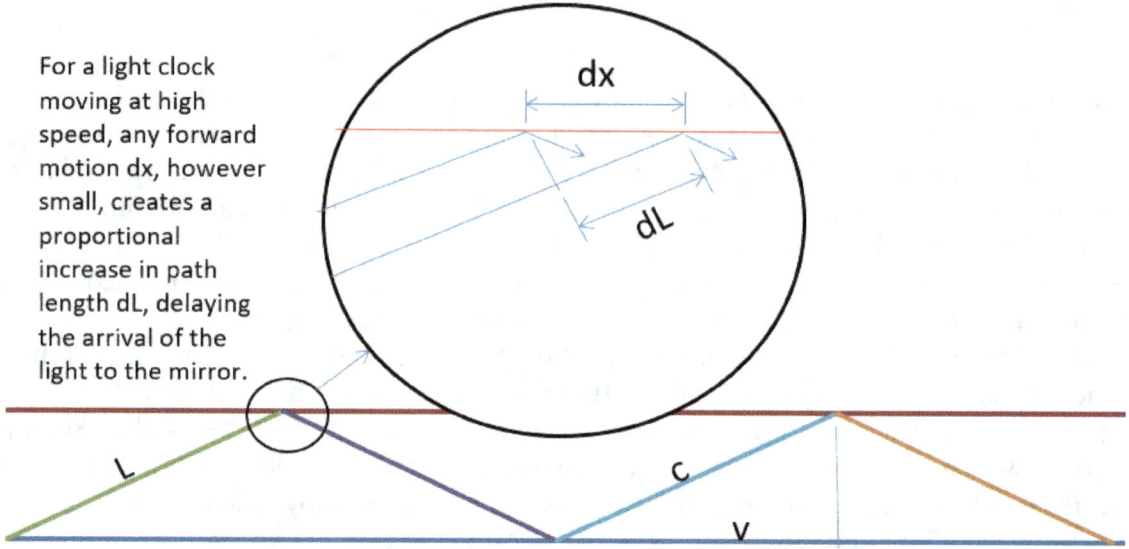

For a light clock moving at high speed, any forward motion dx, however small, creates a proportional increase in path length dL, delaying the arrival of the light to the mirror.

Note: $dt/T = dL/L$, and $dL/dx = v/c$. This matches Lorentz's equations.

These billions of tiny extra distances which we are adding to the length of the light path make the forward-rolling clock run just a tiny bit slower than our stationary clock. And this extra slowness results in the clock that is rolled forward in the ship running a little behind the 'stationary' clock which we have not rolled forward. By rolling a clock forward, it has automatically been set back just enough to correspond to Einstein's time-skewing synchronization procedure which at first seemed so suspect. We also find that if we roll another clock from amidships toward the aft of the ship, we will be shortening the zig-zag path of its light beam, causing it to run a tad faster than the mid-ship clock, and so setting it forward in time just the right amount to maintain Einstein synchronization. And, it's no surprise that the forward clock and aft clock will also be synchronized in the Einstein sense. Using any or all of these clocks to measure the one-way speed of light between them will, as always, result in the same numerical value of the speed of light.

What these thought experiments show is that there is more to Einstein's notion of simultaneity in an inertial reference frame than a self-fulfilling prophesy. This time setback or set-forward must not only happen to clocks that are moved around in a reference frame moving through space, it must also happen to all physical objects that are moved around in the moving frame. It is this effect that causes Einstein's Principle of Relativity to be true. These three effects of time dilation, length contraction and time skewing act in concert to make the Laws of Physics the same for all observers moving with at constant velocity, regardless of their speed relative to space. But instead of starting with this as a fundamental principle, we have started with more primitive concepts: that all matter is composed of nothing but energy, and that all energy travels through space at a constant speed c - and we have arrived at the same place. I don't think this diminishes the Principle of Relativity, I think it enhances the understanding of it and gives it more physical meaning.

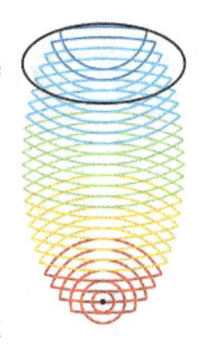

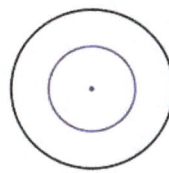

Don't want to look suspicious or I'm done -
But I really think this rhythm is the only one!
So, get this secret out
Deep, distant, and pure…

- Rob Dickenson and Brian Futter, (Catherine Wheel), <u>Kill Rhythm</u>

Amazingly, what our constructive model tells us is that because all matter is composed of energy that at the micro level always travels at the speed of light, sentient beings with measuring instruments will perceive their own environment to be identical to what they would expect if they were not moving through space. For if matter was in fact composed of hard little spherical nuggets of stuff, and clocks and personal time did not slow down because of motion through space, and lengths were not shortened in the direction of motion by flattening of spherical trapped energy patterns, and clocks did not wind forward or backward to maintain Einstein synchronization, then we would *only* expect to measure the speed of light as a constant in all directions if we were NOT moving through space at all. Yet we must be moving through space, at least as much as our speed around the sun indicates and in all likelihood at a much higher speed, as we track along with the sun in its rotation about the center of the Milky Way, and as we track with the center of the Milky Way as it moves through its local galactic cluster. The Michelson-Morley instruments were sensitive enough to detect a velocity as small as the Earth's orbital speed around the sun alone if their assumptions had been correct. But they assumed the hard-nugget structure, constant clock tic rate and universally synchronized clock conceptual models, so they were baffled by their result. They showed that the speed of light is measured to be the same in all directions, but they had no idea why. In fact, they highly expected otherwise.

Despite this remarkable reality, it's amazing to note that when this fact emerged experimentally just before the beginning of the 20th century, no one, not even Einstein, could imagine a constructive model to explain it physically. And worse, when new facts came in the early part of the 20th century about the dual wave-particle nature of both light and matter, no one looked back to try and develop such a model. At least, no one was successful at convincing the physics establishment that such a model was conceivable and reasonable. Instead, Physics moved in the direction of mathematical abstraction. The relationships between different frames which Einstein developed were perceived not only as a practical way to solve problems and make predictions; the new mathematical abstraction called space-time was increasingly viewed as the best representation of reality. Physics became divorced from the philosophy of physical interpretation and fell in love with its new mistress, mathematics.

All my talk so far about relativity has been limited to interpretation of a special case which Einstein and Physics calls 'Special Relativity.' Special Relativity (SR) simply deals with situations in which gravity and acceleration are not at issue. The space I've been talking about so far has been assumed to be removed from gravitational fields, and velocities of reference frames have been assumed to be constant. This was how Einstein began also, and after developing Special Relativity, he spent over a decade bringing gravity and acceleration into the picture. His theory which included gravity and acceleration is called General Relativity (GR.) The mathematical abstraction of physics reached new heights with Einstein's development of GR, and abstract mathematics remains today as the essential way to practice physics.

In the second section of this book, we have more topics that must be covered in the realm of Special Relativity (SR), and I will leave consideration of gravity to a second volume. However, I think it's safe to say that Einstein's development of GR cemented in place the way Special Relativity was understood. Other physicists weren't inclined to look back, they were trying to *keep up*, and look ahead. It was GR that made Einstein famous, a cultural superstar, a household name. This fame came mainly from GR's successful prediction of the bending of starlight by gravity, as measured by Arthur Eddington on the occasion of a solar eclipse in 1919. Surely, with that success, the new mathematical physics was proven to be the wave of the future, and physical interpretation was relegated to a status of 'un-necessary.'

One might also think that modern physicists, having a century's worth of new information, have re-evaluated Special Relativity and that on some level they 'get it' – yes, maybe they get it, but they just don't think it's worth going back to explain the physical underpinnings. Maybe they are too busy to explain it with a physical constructive model in light of the new information. Sadly, this seems not to be the case, because many top-notch physicists have taken the time to write introductory textbooks about relativity, and they all are still stuck in the Einstein groove. All you have to do is pick up any standard introductory book on Special Relativity to see what I'm talking about. It's not that they are wrong. It's just that there is something missing. A constructive model is missing. And without that physical model, I think relativity is much more difficult for people to understand.

So, the first and most important part of the job is done. We have started with a reasonable constructive model and ended up with a reasonable physical explanation of why the one-way speed of light is measured to be the same regardless of the speed of the reference frame that is measuring it. Now we have finally reached Einstein's starting point. We're in a better position to follow his development of relativity forward, but if things seem to get crazy or confusing, we have something to fall back on. For with a constructive model in our heads, we have an understanding of what's really going on which isn't normally available when studying relativity. Now we have a real shot at having it all make sense and feel right. We have a real shot at de-mystifying relativity. We might even have a shot at restructuring the debate about the nature of truth.

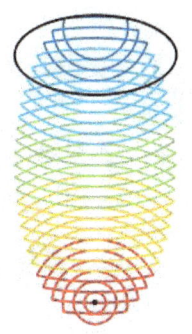

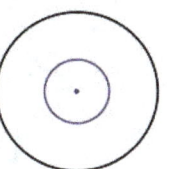

SPECIAL RELATIVITY - PART 2: APPLICATIONS

The light that you will never see
It shines inside; you can't take that from me.
The truth is never far behind, you've kept it hidden well
Hope I live to tell the secrets I have learned
Till then they will burn inside of me.
   - Madonna & Leonard Patrick, <u>Live to Tell</u>

CHAPTER 7 – THE PRICE OF ABSTRACTION

In Einstein's original 1905 paper on Special Relativity, he came tantalizingly close to saying that light travels at a fixed speed through space. His initial statement of his postulates sounded as if that was what he was saying. But very soon thereafter, in the same paper, he changed his wording slightly. He set up a situation in which one observer was moving at a constant speed relative to another observer and said that light traveled at speed c in the frame of each observer. By the time he finished the derivation of the equations of Special Relativity, the concept of space as a physical entity was virtually eliminated. All that mattered, he said, was the relative speed of the two reference frames. The manner in which light traveled through space was no longer a topic. Space itself was eliminated from the discussion.

Einstein said that there is no 'preferred' reference frame. In other words, if there happened to be a reference frame which was 'still' in space, such as we saw in bug swarm plot 1, where v=0, Einstein said there would be nothing special about it. Think about this. Although a v=0 frame in space seems special in that it represents the maximum possible time rate, perfectly spherical atomic structures, and clocks that were absolutely synchronized, with no set-forward time skewing, such a frame was not special in Einstein's mind, because to anyone in any frame, this special state is undetectable. Behaviorally, perceptually, visually, it would be impossible to tell if you were in a v=0 frame or not.

However, in my mind, even if an observer living in such a v=0 reference frame cannot claim a special status in terms of claiming a superior truth of their measurements (because if one is in it, one cannot know that he is in it,) this does not mean that such a v=0 cannot be imagined conceptually. It is through the activity of imagining a reference frame that is fixed in space that one can understand the phenomena of relativity physically. It was Einstein's 'free will' decision to ignore the concept of a v=0 reference frame that has led to a mathematically efficient but abstract mathematical analysis of events occurring in reference frames which are moving relative to each other.

The price of this mathematical abstraction is a mystification of relativity, and worse, the development of a relativistic philosophy in which truth is relative, not absolute. This notion has been strongly reinforced by writers such as Martin Gardner, who attempted to explain relativity "to the millions." Gardner stated, for example (italics are his):

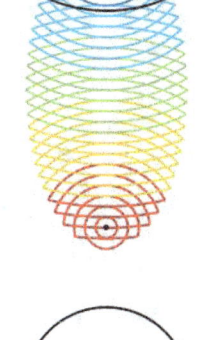

"But the greater the distance between two events, the greater the difficulty of deciding about simultaneity. It is important to understand that this is not just a question of being unable to learn the truth of the matter. *There is no actual truth of the matter.* There is no absolute time throughout the universe by which absolute simultaneity can be measured."

That's an amazing statement. It says that the only truth is a measurable truth - that there can be no other kind. It says that because different measurers will align sequences of events along different timelines, since no definite one-to-one correspondence in time can be determined, there is no actual one-to-one correspondence at all. Since an actual (absolute) truth cannot be measured, there is no actual truth. A few pages later, we see how Gardner leads us to the concept of relative truth. In discussing how two observers measure the other's clocks to be slower than their own, he says:

"The fact that these bewildering changes of length and time are called "apparent" does not mean that there is a "true" length of time which merely "appears" different to different observers. Length and time are relative concepts. They have no meaning apart from the relation of an object to an observer. There is no question of one set of measurements being "true" and another set being "false." Each is true relative to the observer making the measurements, relative to his frame of reference. *There is no way that measurements can be any truer.* In no sense are they optical illusions, to be explained away by a psychologist. They can be recorded on instruments. They do not require a *living* observer."

Gardner's statements, and others which are easily found in relativity books imply that because different observers have different senses of simultaneity, there are multiple truths. This conclusion is reached because science has accepted that measurement is the ultimate truth. They have converted the absolutely true fact that "light is MEASURED to have the same speed in all directions by all observers" into this statement which is not true: "Light travels at the same speed in all directions relative to all observers." Here is how Gardner makes that leap: He first states Einstein's fundamental postulates:

1. There is no way to tell whether an object is at rest or in uniform motion relative to a fixed ether.
2. Regardless of the motion of its source, light always moves through empty space with the same constant speed.

So far, so good. Gardner then says, with no physical explanation:

(The second postulate should not be confused, as it so often is, with the constant speed of light relative to a uniformly moving *observer*. This is a deduction from the postulates.)

Notice that nowhere in *this* statement does Gardner mention measurement. Rather, he seems to say that the speed of light IS constant relative to a uniformly moving observer, and that this fact has been deduced from the fundamental postulates. With our constructive model, we know that waves of light do not really travel at the same speed relative to every observer, but we understand why *measurements* of the speed of light in all directions in a uniformly moving reference frame will yield the same numerical value. The combined effects of slowing atomic processes, non-spherical shortening of structures of matter, and time skewing synchronization of a moving reference frame all produce the inevitable conditions in which measurement of light speed will be identical.

It is this literalism of equating measurements with reality which leads ultimately to the fractured sense of reality which the standard interpretations of relativity leave us feeling. When different participants observe different sequences of events should we conclude that there are two equally valid realities? Or would it be better to understand and remember why this can happen, and state that there is one reality to the sequence of events occurring in the universe, which is not knowable by either observer?

CHAPTER 8 - LADY ON A TRAIN

Let's consider a few ideas related to this time-skewing concept. There is a very interesting quote in one of Einstein's books related to it. He describes his famous 'lady on a train' thought experiment. In this experiment there is an observer on the ground, the embankment he calls it, as a train passes. A lady sits in the middle of the very fast train, at point M. The train is presumed to be traveling at a significant fraction of the speed of light. Just as she passes, two flashes of light occur, which are at equal distances ahead of the lady at point A, and behind her at point B, and therefore also at equal distances from the observer on the embankment at point E.

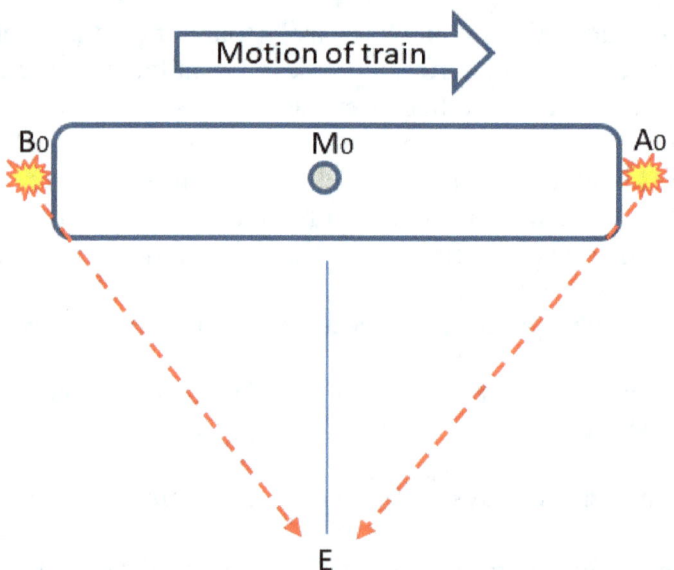

The observer on the ground at E, presumably in a frame that is nearly still in space, sees the two flashes at the same time. But the lady on the very-fast train sees the forward flash from A first, since she is traveling toward the source, reducing its travel distance to her. A small time later she sees the flash from B since she is traveling away from the light source, so the light must travel further to reach her.

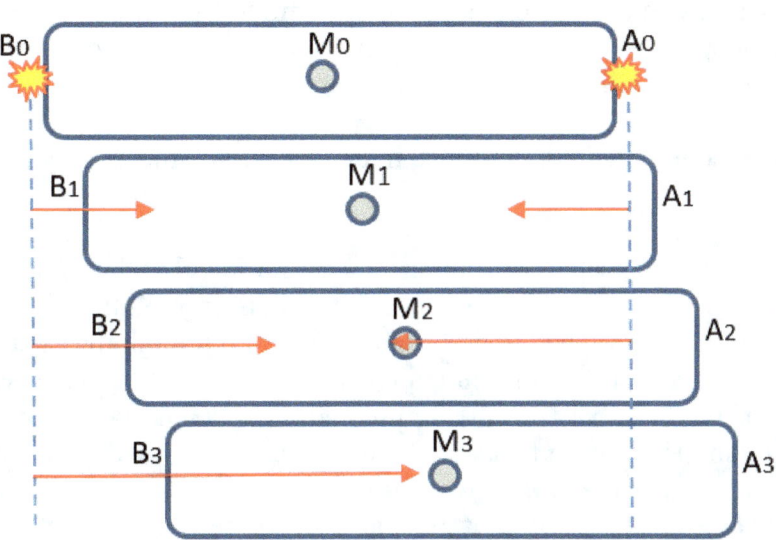
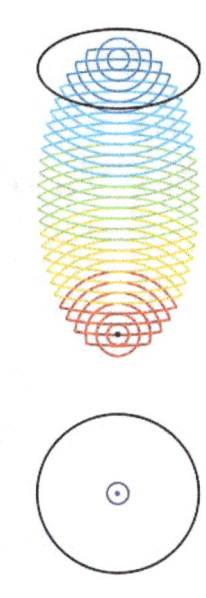

The converse case would be that if the two flashes happened such that the lady saw the two flashes at the same time, then the observer on the ground would see the flash from B occur before the flash from A. This is Einstein's example of the relativity of simultaneity. He argues that the lady on the train must make the assumption that it takes the light the same time to travel from A to M as it takes to travel from B to M. In other words, in her frame of reference (and according to Einstein's clock synchronization procedure), light travels the same speed in all directions. Here is the interesting quote:

"That light requires the same time to travel the distance from A to M as from B to M is in reality neither a supposition nor a hypothesis about the nature of light, but a stipulation which I can make of my own free will in order to arrive at a definition of simultaneity."

This word stipulation is very interesting. A stipulation is an agreement. The Wiktionary says that the Romans broke a straw – a stipula in Latin – to show agreement, indicating that a particular issue would not be contested. Lawyers define a stipulation in the same way – a point that both parties agree will not be argued.

If this book was titled "Deconstructing Relativity" rather than "Reconstructing Relativity," I would be focusing heavily on this famous quotation. Who are the two parties to this agreement? Perhaps something was lost in the translation, but this appears rather to be a decision instead of a stipulation, since there is no dialectic here, there is no opposition party. This was, as I indicated earlier, a brilliant decision which provides a means to establish a rational definition of reference frames in which the one-way speed of light will always be measured to be the same regardless of one's motion through space – a fact which agrees with experimental results. This is the core of the "universal formal principle" which would lead to "assured results." Again, the statement that this is "in reality neither a supposition nor a hypothesis about the nature of light" seems to acknowledge the absence of a constructive model which fully explains the physical phenomena. But my intention here is not to undermine Einstein's relativity, rather my intention is to support its foundation by adding a constructive model which explains why his formal principle is true.

In one relativity book I studied, the Einstein definition of simultaneity is described this way:

"Two travelers in space keep a constant distance, and travel at the same speed in the same direction. They synchronize their watches with the assumption that light is traveling between them at the same speed in both directions, even though they realize that this might not be true, because of their mutual motion through space. They know that the 20 second round trip of light between them might actually be taking 11 seconds in one direction, and 9 in the other, but they set their watches as if it were 10 seconds in each direction, because *that's the best they can do*. Since they have no other means to measure the actual time of each transit of light, they agree that the time for each transit is equal."

These stipulations, these agreements, are very reasonable and practical for anyone operating in an extended frame of reference, and, come to think of it, every frame of reference must, by definition, have extension. Whether the extensions are in light years across space, or only meters in a physics lab, it takes time for light to get from A to B, and there is no "locatable" absolute reference by which to measure one's absolute speed though space. Clearly Michelson and Morley failed to measure the Earth's speed through space with their technique, and they just as famously proved in retrospect that it could not be measured.

CHAPTER 9 – ON SPACETIME

"But now old friends are acting strange
They shake their heads, they say I've changed
Well, something's lost, but something's gained
In living every day"
    - Joni Mitchell, Clouds

After Einstein's conceptual breakthrough and his derivation of the Lorentz equations based on his definition of simultaneity, Physics developed the mathematical abstraction called spacetime. Spacetime is a four-dimensional mathematical construction in which every event which occurs in the universe is conceptually located by 4 co-ordinates: x, y, z, and t. In the classical thinking that preceded relativity, the position in space of an object was defined by x, y, and z coordinates. Although the selection of a reference frame was arbitrary, computations of the distance between any two points in space did not depend on the reference frame that was used. Distance was considered an absolute entity that all observers would agree on. Time, in pre-relativity thinking was also assumed to be absolute and universal, and not dependent on one's reference frame.

In the relativistic mathematics, however, space and time are considered to be intertwined.

Because of the constancy of the measured speed of light, time can be put on the same footing as space in an odd way. The unit of the 'light-year' perfectly demonstrates this principle. A light year is the distance light travels in a year. A distance as measured in any reference frame can be evaluated in terms of time measured in that reference frame; time in a reference frame can be evaluated by the distance light travels in the same measuring system. This works because, as we have seen, in any steadily, linearly moving reference frame, the speed of light is always measured to be the same in all directions. But when thinking about our constructive model, it is important to remember that the key word in this line of thinking is 'measured.'

Spacetime Relativity treats reality as a series of 'events' where every event can be labeled as occurring as a particular (x,y,z,t) coordinate in space-time. With the Newtonian absolute agreement on lengths and time lost, relativity arrived at a new absolute in its space-time mathematical model. This new absolute is called 'proper time.' Proper time is often called the 'distance' between two events in the (x,y,z,t) space-time model, but calling it a distance is a mathematical analogy. Let dx= the measured difference in the x coordinate on an arbitrary reference frame, and dy and dz similarly the difference in y and z measurements between two points. Just as distance between two points in space in Newtonian mechanics is computed as $\sqrt{dx^2 + dy^2 + dz^2}$, in relativity, proper time is computed as $\sqrt{dt^2 - dx^2 - dy^2 - dz^2}$. Note that dt is similarly the difference in the times of two events as measured in a reference frame.

Where Newtonian thinking said that regardless of the reference frame used, the computed distances would be equal and dt would also be equal for any measurers, Special Relativity says that proper time will be equal, but distances and times will not be equal if the frames have a uniform relative speed. But what is proper time? What does it mean? It is the 'personal time' or the time that would be experienced by a person (or measuring apparatus) that traveled such that they were physically present at both events. Well, this is reassuring. At least different observers can agree on some things. Even with clocks that may be running at different rates, and meter sticks of different lengths, all observers in frames moving at constant relative speeds will compute the same proper time between two measured events. And they should, because a traveler who is present at both events measures the elapsed time on her clock between the two events, and she has only one number for that elapsed time on her clock. In the traveler's frame, both events occur at x=0, y=0, and z=0, and the time interval she records (dt) is her proper time.

The formula for proper time is quite interesting. Notice that while dx^2, dy^2 and dz^2 all have positive signs, dt^2 has a negative sign. The result is that if you measure someone or something that travels a relatively small distance over a long time period, their clock will run at about the same rate as yours, but if you measure something traveling a very, very, large distance in a very short time, their clock will appear to you to be running slower than yours. It is said that to travel further through space, is to 'travel less through time.' That's a clever and concise statement, but it doesn't explain how this happens physically. With our constructive model, we have a notion of how this works: experiencing less personal time means stretching out the helix and slowing the rate of the energy cycles that compose all matter.

Where did this formula for proper time come from? It was derived by Einstein's mentor Minkowski from the Einstein/Lorentz transformation equations by noticing the abstract mathematical form of the equations. Minkowski's formula is not only correct, it is a very simple and efficient formulation of the equations of Special Relativity. Working with this 4-dimesional space-time model with all its mathematical elegance opened the door to further abstract mathematical developments such as Einstein's General Relativity. GR probably would not have been possible without this abstract approach. So much was gained, but what seems lost in it is any connection to intuitive explanations of the physical phenomena.

CHAPTER 10 - TWIN PARADOX

Now I think it's time to take on several so-called paradoxes of relativity, and I plan to do this primarily in a qualitative way. What I mean is that we already have the main concepts in hand, and rather than dealing with equations, I'll give you an example with a few numbers that demonstrate the basic principles we've already discussed in an easily digestible way. Working with the Einstein/Lorentz equations is not very difficult for those who wish to follow through quantitatively, but my goal here is physical explanation in terms of the concepts we have been discussing. The first paradox is the famous Twin Paradox, in which a space-traveler returns younger than his/her stay-at-home twin. Along the way, however, it will be convenient to resolve another paradox that I've already alluded to. This is the seemingly paradoxical fact that two observers in high-speed motion relative to each other will BOTH measure the other's time as slower than their own. This is the "time symmetry paradox." After resolving these, we'll get to the slightly more involved explanation of how each observer also measures the other's lengths to be shorter in the direction of relative motion than their own.

To set up this Twin Paradox example, I want to make one approximating assumption which will make this explanation most simple, yet I hasten to add that it is not a necessary assumption for the results of the Twin Paradox to be what they are. Again, Physics is a very specific business, and I want to be clear from the start that I have thought through other scenarios as well. What I'd like to assume for this first run-through is that Earth's speed through space can be considered approximately zero, in contrast to the hypothetical futuristic rocket ship which can get to an extremely high speed such as the familiar .866c. If you will accept this conceptual premise for the first example, I promise I will also address what happens under a very different assumption at the end of this chapter.

I also would suggest that although the approximation of Earth having zero speed through space is certainly not exactly true, it may not be a bad approximation, when one considers that the speed of light is 186,000 miles per second, which means sunlight reflected from the moon reaches us in about 1.2 seconds. Since Apollo astronauts took roughly three days to get to the moon (though not at constant speed) we clearly recognize that an example with rocket v=.866c is extremely un-do-able with current technology. At this speed, a trip to the moon would take about 1.5 seconds, not the 3 days it took Apollo ships. The nearest stars to Earth are several light-years away, but we observe that as years go by, our motion relative to these stars must be quite small, or we would notice significantly rapid shifts in their positions and the distances to them, which we do not observe.

All right then, for the purposes of discussion in terms of our constructive model, we will first assume that a stay-at-home twin is approximately not traveling through space at all, such that all atomic processes, clocks, and heartbeats run very close to the maximum actual rate possible in the universe. Along with this, for the purpose of this 'thought experiment' let's assume that Earth has a number of remote 'way-stations' in space that all have identical clocks and calendars that are synchronized with Earth's clocks and calendars. For example, there is a station a light year away from Earth such that when Earth sends out a radio signal on New Year's Midnight, January 1, 2050, the station receives this signal when its clock reads New Year's Midnight 2051, and when this station immediately sends a return signal, Earth gets this return signal at New Year's midnight, 2052. Earth has many similar stations at further distances, and they are all synchronized in this fashion, which is indeed the only practical and sensible way to do it.

Note that with our primary assumption of Earth not moving ($v=0$) through space, Earth's clocks and calendars are 'truly synchronized.' Earth and all its way-stations form a reference frame which is in a 'bug swarm 1' state, in which there is no time skewing. Although they can't really know it, because they can't really KNOW that they are in a $v=0$ condition, we have assumed that for this discussion, so under our assumption, when Earth and people living at these remote way-stations celebrate New Year's Eve, 2051, they actually all 'pop the cork' at the same time in State Universe. In other words, the Universe would never exist in a state in which one of the Earth stations had already popped the cork, and some of the others had not.

Now we introduce the super-ultrafast .866c rocket ship traveling away from Earth on a multi-year trip. Exactly how many years the trip lasts really depends on whether you are on the ship, or on Earth. Remember that for $v=.866c$, Gamma = 2, which means that the ship and all its sub-atomic structures, so therefore its inhabitants and clocks, are operating at half-speed. Their atoms are flattened, and their time is skewed as we see in bug-swarm plot #3. In order to explain the phenomena at hand, we need to construct a reference frame for the traveler which is similar to the reference frame which Earth has, because we want to see how they take measure each other's reality, in particular for this example, the rates at which they perceive each other's clocks to run. Remember, we have to explain two curious effects, which combined together seem even more strange: They both perceive each other's clocks to be running slower than their own, and yet, the traveler ends up younger at the end of the trip, a fact which they both agree on when the traveler returns.

For the traveler to have a frame of reference similar to Earth's frame of reference, we need to imagine that the traveler and his/her rocket has a series of way-stations of its own, which are spaced out at great distances from the ship yet travel along with the ship in the same direction, at the same speed (.866c) through space. We assume that this 'wagon train' of traveling way-stations has been moving together for a very long time, and so it has been able to synchronize its clocks per Einstein's synchronization rule. Unlike the v=0 Earth reference frame which has no time skewing, the v=.866c reference frame has very significant time skewing. In other words, the traveler's wagon train clocks <u>do not</u> read the same time at the same moment in State Universe. If we imagine the outward leg of our traveler's trip to be moving to the right, then the bug swarm plot #3 is the appropriate constructive model, and we remember that in order for the traveler's clocks to be synchronized in his frame, the clocks on mini-ships that physically lead this wagon train must be set back in time relative to his ship's clock, and clocks on mini-ships that follow the ship in the wagon train must set their clocks forward in time relative to the traveler's ship clock. And the further a 'wagon train' clock is away from the traveling twin, the more its clock is set forward or back.

This time skewing is the key to the paradox in which both observers measure the other's clocks as moving slower than their own. The next chapter will give a detailed example of this, but this is also a good moment to remind you of the philosophical difference between viewing this situation with Einstein's Formal Principle of Relativity, and viewing the situation with a physical constructive model. With the Principle of Relativity, and the Lorentz formulas that Einstein derived from it, both reference frames measure the other's clocks to be slower than their own, but there is no understanding of how this happens physically. The symmetry is in the mathematics, but the mathematics was derived from a postulate that was never explained physically, so this gives the sense of mystery that earns the title 'paradox.' The result seems illogical because indeed there *was* a gap in the logic: when Einstein used the proven experimental fact that the speed of light is measured to be the same by all moving observers as his formal principle, he didn't have a constructive model that explained the underlying physical phenomena.

With a constructive model in mind, we can imagine what is really happening, such that this 'paradox' does not seem so mysterious. In the scenario presented, we have two very different things going on. The Earth frame measures the traveler's clock to be going slower that its own because the traveler's clock really IS going slower than the Earth clocks, because of the traveler's rapid motion through space, and the Earth clocks are truly synchronized in the sense of State Universe, not just in the Einstein sense. On the other side, the Traveler measures Earth's clock as going slower than its own for an entirely different reason. The traveling frame measures the Earth's clock as slower because of the time skewing of the traveler's set of way-station clocks.

So, let's continue to consider the twin paradox with these facts in mind. We would start, of course, with the Earth clock and the ship clock both reading zero. As the spaceship moves off into space at .866c, his wagon-train of trailing mini-ships begins to pass Earth in a steady stream. Each wagon-train ship that passes Earth is successively farther in distance behind the traveler's spaceship than the prior wagon-train ship that passed Earth. Each successive wagon-train therefore has its clock set forward in time more than the previous wagon-train ship. Although these wagon-train clocks all run at the same rate (half-speed compared to Earth), they are not synchronized in the 'State Universe' sense; they are only synchronized in the Einstein simultaneity sense for the moving frame. As each wagon-train clock passes Earth, it compares its own clock to a clock it sees on Earth, and each wagon-train station records that its own clock is ahead of Earth's clock. And as each successive wagon-train ship passes Earth, each finds that Earth's clock is farther behind its own clock than the prior wagon-train ship's assessment. Thus, the spaceship's 'frame of reference,' which is its only means of measuring, records that Earth's time is running slower than its own time.

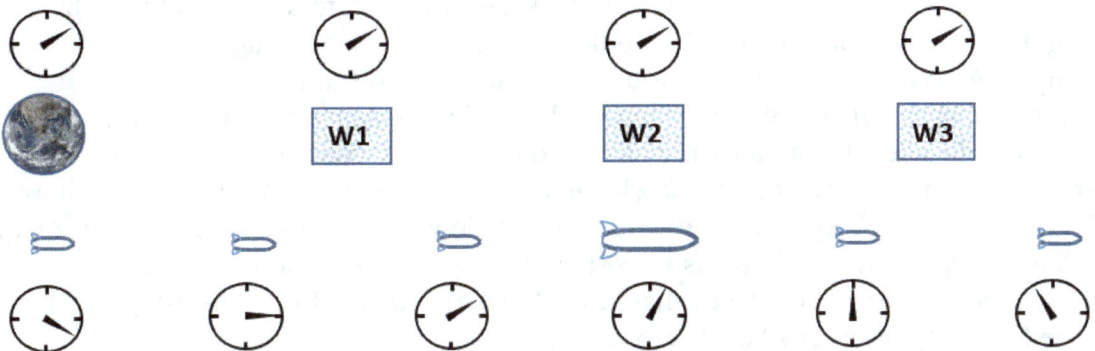

Earth and its way-station clocks, (all at v=0) synchronized in the Earth frame and in State Universe.

The traveler's ship and its moving wagon train, (all at v=.866c) synchronized in the Ship's frame. These moving clocks all run at half speed compared to Earth, and they have significant time skewing.

To give simple numbers to this as an example, in the scenario described with v=.866c, when the Earth clock/calendar registers two years after ship departure, the traveling ship, which exists in a slowed-down state, has only experienced one year on its calendar. An Earth way station clock located near the ship at this moment would also read 2 years on its clock calendar at this moment, and it would see that the clock calendar on the ship is at its 1-year mark. But the ship's trailing Einstein-synchronized wagon-train ship which is passing Earth at this same moment in the State of the Universe, has its clock calendar set forward so that it is reading 4 years after the date of the Ship's departure, because this wagon-train clock is, and ever since synchronization with the traveler's spaceship it has been, set forward in the time-skewing sense we have discussed. It is this time-skewing in the moving frame which produces the symmetry of measurements. At 4 Earth years, the spaceship clock reads 2 years, but the space-ship frame-synchronized wagon-train clock passing Earth at this moment reads 8 years. At 6 Earth years, the spaceship clock reads 3 years, but the space-ship frame-synchronized wagon-train clock passing Earth at this moment reads 12 years. And so on.

If you're like me, you might read this last paragraph, and study the graphic above it and say… wait a minute! When one of these waystation clocks looks at the traveler's main ship and measures that the ship's clock is running slow, can't the person on the ship also look at the waystation clock and AGREE that the ship's clock is running more slowly than the Earth's set of synchronized clocks? And can't the folks on Earth also monitor the times on each passing wagon-train clock, and notice that Earth clocks are running slower than the traveling wagon-train clocks? After all, we might assume there are intelligent people on all these way-stations and wagon-train mini-ships. Well, yes, they could, but to do so would be a sort of Subject-Object role reversal in this thought experiment.

When Science says that a measurement is made, the Subject (observer) uses instruments in its own reference frame to measure properties of the (observed) Object. Thus, Subject Earth may use its set of synchronized remote sensing waystations in its reference frame to observe and measure the clock rate of its Object: *the traveler's ship*. And the ship, as a Subject, may use its set of synchronized remote sensors on its traveling wagon-train reference frame to observe the clock rate of its Object: Earth. For the ship to draw conclusions by reading Earth's waystation clocks, or for Earth to draw conclusion by reading the ship's wagon-train clocks would require *a priori* knowledge – in other words, knowing in advance that these waystation clocks and wagon-train clocks are synchronized in *the other's* reference frame. While in this thought experiment, we have 'people on both sides' taking measure of each other, an observed Object doesn't generally have this sort of a priori knowledge of the Subject that is measuring it or how its Subject's measuring apparatus are arranged, let alone the consciousness to interpret it. This "no a priori knowledge" rule is important in fully understanding how relativity shows that both observers conclude that the (Object's) "other" clock runs more slowly than the (Subject) observer's "own" clock.

Anyone who's read a few books about relativity probably knows that there is some controversy about the Twin Paradox that involves the turn-around of the space ship. In the early years of the 20th century, lacking a constructive model of how relativity worked physically, there was a lot of confusion about the 'symmetry of measurement' which we have just discussed. The Twin Paradox was doubted by some who saw it as incompatible with the 'symmetry of measurements' phenomenon. They argued that if relative motion between two observers is all that matters according to Einstein, there would be no way to determine if the spaceship was actually moving through space, and Earth sitting still, or Earth was moving through space and the spaceship was sitting still. Given this uncertainty, they argued: why is it that the spaceship twin must be the one who ends up younger when he/she is re-united with their Earth-bound twin?

Amazingly, Einstein answered this challenge in 1915 by means of his General Relativity theory. According to this defense of the Twin Paradox, it was the acceleration which the spaceship would undergo during the turn-around which causes the eventual age discrepancy of the twins at the end of the story. The mathematics of this 1915 General Relativity explanation was much more complicated than the original Special Relativity explanation. Einstein pointed out that the Special Theory of Relativity only applied to reference frames moving at fixed speeds without acceleration with respect to each other, so the Twin Paradox could not be fully explained with Special Relativity alone. Clocks run slower in a gravitational field, he said, and the acceleration required to stop the ship and reverse its direction is just like gravity. I laughed out loud when I read about this. So… he used the complicated General Theory which was built on the simple Special Theory to explain the Simple Theory! Now the aging difference was explained as occurring only in the acceleration phase??

Since then, others have argued that the whole issue of acceleration in the Twin Paradox story can be skirted by introducing a second spaceship which is headed toward Earth at the same high rate of speed as the outbound ship. At some point, these ships cross paths, and the ships exchange *information* about their clock times. In this scenario, no acceleration occurs, yet the sum of the travel times (per their day counts on their calendars) of the outbound twin and the inbound traveler is still less than the Earth time accumulated between the departure of the first ship and the return of the other. Since we are clearly hypothesizing about the future, it seems we could also propose a version of the Twin Paradox in which the outbound twin teleports (as in *Star Trek*) from his ship to the inbound ship as they pass. This is again simply a weightless transfer of *information* from one ship to the other: in teleportation, it is said, the information about the structure of the original object can only be gained by destroying the original structure, the original outbound twin in this case. The teleportation receiver on the inbound ship would use the received information to create a new twin in the inbound ship identical in form to the destroyed outbound twin, thus re-creating a human being on the inbound ship of the same age as the one that was destroyed on the outbound trip. No acceleration, there!

In any case, our constructive model tells us that the change in aging rates is caused steadily and continually throughout both legs of the trip by the motion of the traveler through space. I want to finish the example we started by giving numbers for our example scenario's second leg, and after that, as promised, I will come back again to discuss scenarios in which Earth is not sitting still at v=0 in space. But for now, assuming that Earth is at v=0, what is really going on during the second leg of the Twin Paradox? Let's assume we use teleportation to transfer the traveling twin into the inbound ship without material acceleration. Let's suppose this transfer happened after the traveler has counted 3 years of life on the outbound ship, and again, let's assume an inbound speed of .866c, which still corresponds to gamma = 2.

The traveler is now in a different reference frame, and the time skewing is reversed. That is, if this inbound ship also has a wagon-train fleet of way-stations traveling along at the same speed ahead of it, and each wagon-train station has an Einstein-synchronized clock onboard, these way-stations which are ahead of the traveler's ship will have clocks that are set *back* in time in the 'State Universe' sense relative to the inbound traveler's ship clock. That is, they will be synchronized as a functional reference frame, and they will run at the same clock rates (half speed because gamma = 2), but they will not 'really' be showing the same clock times at the same moment in the State of the Universe sense. Again, we assume these inbound clocks have had a very long lead time to get synchronized, and to complete our scenario, let's say that the inbound clocks will reference all times such that the inbound ship time when the traveler is transported onto it is labeled exactly 'three years.' In this way, we can continue accumulated times which were established in the outbound leg of the trip.

We will find again, that on the inbound leg, both observers still measure the other's time to be running at half of their own speed. Again, it is the case that the Earth frame comes to this conclusion because the Earth's clocks have no time skewing, and the spaceship's clock is indeed running at half speed, and the inbound traveler measures the Earth clocks as running at half speed relative to it because of the time-skewing associated with the traveling clock's reference frame. I will make a simple table below which will summarize both outbound and inbound clock/calendar readings, and then we can discuss it more.

Earth Clock & Way-stations	Ship Clock	Ship Way-station Clock by Earth
0	0	(0)
½ year	¼ year	1 year
1 year	½ year	2 years
1.5 year	¾ year	3 years
2 years	1 year	4 years
3 years	1.5 years	6 years
4 years	2 years	8 years
5 years	2.5 years	10 years
6 years	3 years	12 years
- - - - - - - - - - - teleportation of traveler to inbound ship - - - - - - - -		
6 years	3 years	-6 years
7 years	3.5 years	-4 years
8 years	4 years	-2 years
9 years	4.5 years	0 years
10 years	5 years	2 years
10.5 years	5.25 years	3 years
11 years	5.5 years	4 years
11.5 years	5.75 years	5 years
12 years	6 years	(6 years)

After 12 Earth years, the traveler has returned, only 6 years older than when she left. Where did the traveler's time go? It was not lost in some discontinuity at the turn-around. The only discontinuity occurring at turn-around is the status of the time-skewing from a trailing (clocks set forward) wagon train way-stations to a time skewing of equal magnitude but opposite direction for a leading (clocks set back) wagon train way-stations. Acceleration did not cause the traveler's clock to slow down. The traveler's clock, breathing, heartbeat, neural firing rate, and every atomic and chemical process simply ran at a slower rate compared to the Earth-bound rate for the entire duration of the trip. In a State Model of the Universe, the traveler was alive and present in each of the 12 years experienced by the Earth. The traveler however experienced a number of experiential cycles consistent with only 6 years of normal life.

You may have noticed that in discussing the twin paradox, I have talked a lot about times, but I have not said much about the distances involved. That is because I wanted to delay the discussion of the paradox of symmetry of length measurements for a bit. But first, let's go back to discuss scenarios in which the Earth is not assumed to be stationary in space. When both observers (on Earth and in the rocket ship) are moving relative to space, Einstein's Lorentz Transformations equations correctly describe the differences in measurements between the two reference frames. Einstein is completely right that the relative motion between frames is all that is needed to determine these relationships. The position of absolute 'still space' is irrelevant.

So, let's go back to the loose end that has been left dangling in this 'Twin Paradox' discussion. This loose end pertains to those who questioned the asymmetry of the Twin Paradox in light of the measurement symmetries. The argument went: If we can't know where space is, who is to say that the rocket ship is moving through it faster than Earth? Suppose Earth was somehow already moving at .866c through space (let's say leftward), and a rocket blasts off at .866c relative to Earth (rightward), such that the rocket ship is the entity that is really at v=0 relative to space. Wouldn't the twin on Earth then experience less aging, and the twin in the rocket ship age faster? The answer is yes, on the 'outward leg' of the trip. But if it is the rocket traveler, rather than Earth, that 'turns around' (or again, transfers to a different ship via teleportation if we wish to avoid the acceleration issue) then the ship traveler must travel through space much faster than the Earth is moving through space on the return leg in order to get home, and in the process, the rocket traveler will still be the younger one at the reunion party on Earth.

And the numbers will be the same, if, that is, the rocket can achieve a return speed of .866c relative to Earth on the return trip. The questions arise: "Is it possible to catch up to Earth at a rate of .866c relative to Earth when Earth is moving .866c relative to space? Wouldn't you have to go faster than c to do that? Isn't that impossible?" The answers are yes, it is possible, and no, you do not have to exceed the speed c through space, and yes, it is impossible to go faster than c through space. Also, no observer in a constant-speed reference frame will ever measure another object as moving faster than the speed of light relative to itself. Even if two ships are approaching each other, and both are traveling at .866c relative to space, they will measure a relative speed less than c.

Recall that the simple addition of velocities which we are familiar with from Newtonian mechanics is only an approximation applicable to low speed objects, and at high relativistic speeds, where gamma is measurably larger than 1, relativity gives us a new formula for adding relative velocities.

For calculation of co-linear velocities,

Where Galilean / Newtonian mechanics gave us:
$$s = v + u$$

Special Relativity gives us: $$s = \frac{(v+u)}{\left(1+\frac{v*u}{c^2}\right)}$$

Where: v is a velocity of an object A relative to frame B,
u is the velocity of frame B relative to frame C, and
s is the velocity of the object A relative to frame C.

For our problem, v = vel. of Earth rel. to space = .866c
u = vel. of Ship rel. to Earth = .866c
So, s = vel. of ship relative to space = 1.732c / (1+.75) = .9897c
At this speed, gamma = 7.

So, in the scenario we are considering, Earth is moving at .866c (leftward, say) and the ship is at v=0 on the first leg of the trip, and then in the second leg, the ship moves through space (leftward) at .9897c to return to Earth, while the Earth's speed remains unchanged. If we work the numbers, we will find that:

Earth time runs at half maximum cycle rate (gamma=2) during the entire trip, while the ship runs at maximum rate (gamma = 1) during the first leg, and then runs seven times slower than maximum (gamma = 7) during the ultra-high-speed chase to return to Earth. Note that as before, the ship turns around (or, the traveler transfers to the leftward moving ship) when the traveler's calendar/clock reaches 3 years.

First the table, then more discussion:

Earth Clock	Earth WS Clock by Ship	Ship Clock	Ship WS Clock by Earth
0	(0)	0	(0)
½ year	2 years	1 year	1 year
1 year	4 years	2 years	2 years
1.5 year	6 years	3 years	3 years
-------------	Teleportation transfer of traveler to Earth-bound ship	----------	
1.5 year	6 years	3 years	-15 years
2 years	6+2/7 years	3+1/7 years	-14 years
2.5 years	6+4/7 years	3+2/7 years	-13 years
3 years	6+6/7 years	3+3/7 years	-12 years
4 years	7+3/7 years	3+5/7 years	-10 years
5 years	8 years	4 years	-8 years
8.5 years	10 years	5 years	-1 year
9 years	10+2/7 years	5+1/7 years	0 years
10 years	10+6/7 years	5+3/7 years	2 years
10.5 years	11+1/7 years	5+4/7 years	3 years
11 years	11+3/7 years	5+5/7 years	4 years
11.5 years	11+5/7 years	5+6/7 years	5 years
12 years	(12) years	6 years	(6) years

Notice several interesting things in the above table. Although the relationship with space is different than in our first charted example, it is still true that the relative velocities of the Earth and the ship are .866c (gamma=2) in both scenarios, and the chart shows that both observers' way stations (Einstein-synchronized reference frame clocks) measure the other observer's clocks as running at half the rate of their own clocks on both legs.

Also, the aging difference after the round trip is the same as in case 1, i.e. the ship traveler ages half as much as the Earth twin. Again, this is consistent with Einstein's assertion that only the relative velocities are important. Although Einstein concluded that the notion of absolute space is moot, we again find some value in considering it for the purposes of understanding what is physically happening.

CHAPTER 11 - SYMMETRY PARADOXES

Finally, let's tackle all of the symmetry paradoxes mentioned earlier. We discussed the symmetric time dilation paradox in the last chapter, but now let's consider the symmetric length contraction paradox. In Special Relativity, when two observers move at constant linear speed to each other, they both will measure objects in the other observer's structure to be shorter (in the direction of mutual motion) than objects in their own frames. In other words, at v=.866c, (gamma = 2,) A will measure B's meter stick to be a half meter long, and B will measure A's meter stick to be a half meter long if the meter sticks are aligned with the direction of relative motion between the ships. This seems absurd, until you think carefully about how you measure the length of something that is moving relative to you. You can't just lay your meter stick against the moving meter stick as it goes by. This would not only be extremely impractical at high speeds, it would technically not be measuring the length of the object in your reference frame, because your meter stick would no longer be fixed in your own reference frame. To measure the length of a moving object, what you have to do is record the position *as measured in your frame* of the two ends of the moving object at the same time, that is, *at the same time in your reference frame*. And now you can probably imagine how this "paradox" works.

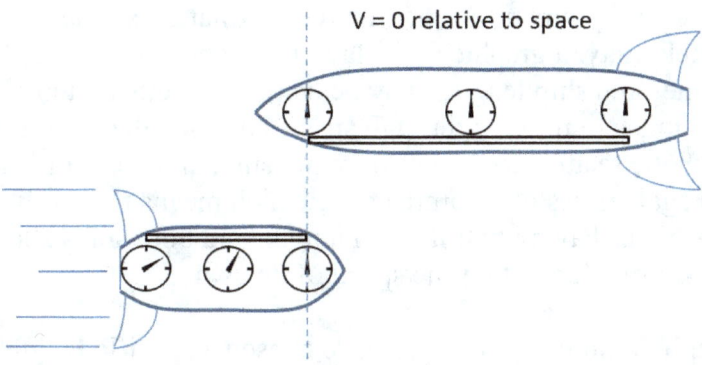

Rather than measuring a meter stick, let's have two identical rocket ships measure the length of each other as they pass. Each ship has a series of synchronized clocks all along the length of their ships. Although it can be shown that this result happens regardless of the 'absolute speed' of either ship through space, let's for the sake of this example consider the case in which one ship is still (v=0) in space, and the other ship at a high speed (again, let's use v=.866c, gamma = 2) relative to it. The v=0 ship will have 'truly synchronized' clocks, but the moving ship's clocks will be synchronized per Einstein, in other words, with time skewing. Its forward clocks will be set back in absolute time, and its rearward clocks will be set forward in absolute time.

The v=0 ship will measure the other ship as being half its own length, because every atom in the moving ship is shortened by a factor of two as shown in the bug swarm metaphor. The v=0 ship will do this by noting that a pair of clocks in its own reference frame set which are half of its own ship length apart recorded the passing of the front and the rear of the other ship at the same recorded time on these two clocks.

Although the moving ship is actually shorter in the absolute sense, when it measures the spatial position of two clocks (in its own reference frame) that recorded the passing of the front and the rear of the other ship *at the same time in its own clock system*, it will also find that these two clocks are separated by half of its own ship length. This happens because the moving ship's clocks are time-skewed.

There is symmetry of measurement, although the physical reasons for the two measurements are quite different in this case. In the more general case where both ships are moving through space, the same symmetry of measurement will exist, although both ships will be shortened in the absolute sense to some degree, and both will have clock systems which are somewhat skewed in the absolute sense. Again, Einstein is right - if your concern is the reality of the measurements, the only thing that matters is the relative motion of the ships. But without a constructive model, it's not very easy to comprehend how this works physically.

DILATION AND CONTRACTION SYMMETRIES – AN EXAMPLE

To demonstrate how the symmetry of measured time dilation and length contraction works in Special Relativity, a graphical display of a quantitative example with some numerical clock times and ship lengths may be helpful. In this thought experiment, let's suppose that one ship is not moving through space, and the other ship is moving at very high speed of $v = .866c$ relative to the non-moving ship, and therefore also relative to space. This speed again gives us a gamma of 2, which means that both ships should measure the other ship as having half its own length, and both ships should measure the other ship's clocks as running at half the speed of its own.

For numerical simplicity in the graphics, I have chosen a specific length for these ships and a convenient time unit for their clocks. The ships are identical and their lengths (as measured at their time of construction) are both equal to 200 light-nanoseconds, that is, the distance that light travels in $200 \times 10^{-9} = 2e\text{-}7$ seconds. A light-nanosecond is the distance light travels in a nanosecond. In more familiar units, 200 light-nanoseconds equals a little less than 60 meters. In the graphics, I show times on three clocks on each ship, one at the forward tip, one amidships, and one at the extreme rear of the ship. The units of time shown on these clocks are 1×10^{-7} (=1e-7) seconds. In other words, it takes these clocks 100 nanoseconds (1e-7 seconds) to register 1.0 unit on their clocks. This means that the ships will measure that it takes exactly 2 units of time on their clocks for light to travel from one end of their ship to the other end, or 1 unit of time from amidships to either end. At this high relative speed of .866c, the ships are not near each other for very long, so these clocks must be very accurate and precise to make such measurements. For the purposes of this thought experiment therefore, we will "make it so..." as they say.

PRELIMINARY STEP: ESTABLISHING TIME SKEWING IN CLOCKS

Before showing how the 'symmetry paradoxes' play out, we need to firmly and quantitatively establish the clock offsets that will be key to understanding these results. For the thought experiment just described, clearly one of the ships will have no time skewing because it will be assumed to be not moving through space. All the clocks on this ship will therefore not only run at the same rate, they will read the same time 'at the same time' in State Universe. There will be no moment in the state of the universe in which any of these clocks will be offset in their time readings from each other.

For the ship that is moving through space (and therefore moving relative to the 'non-moving ship') clocks at different locations will run at the same rate as each other, but they will be offset from each other in 'State Universe.' These clocks on the moving ship will, of course, be synchronized in its own reference frame, per the Einstein synchronization concept, which is the only practical way to establish a frame of reference from which measurements can be made. The following graphics should clearly establish these clock offsets which will be employed in the 'paradox' solutions that will follow.

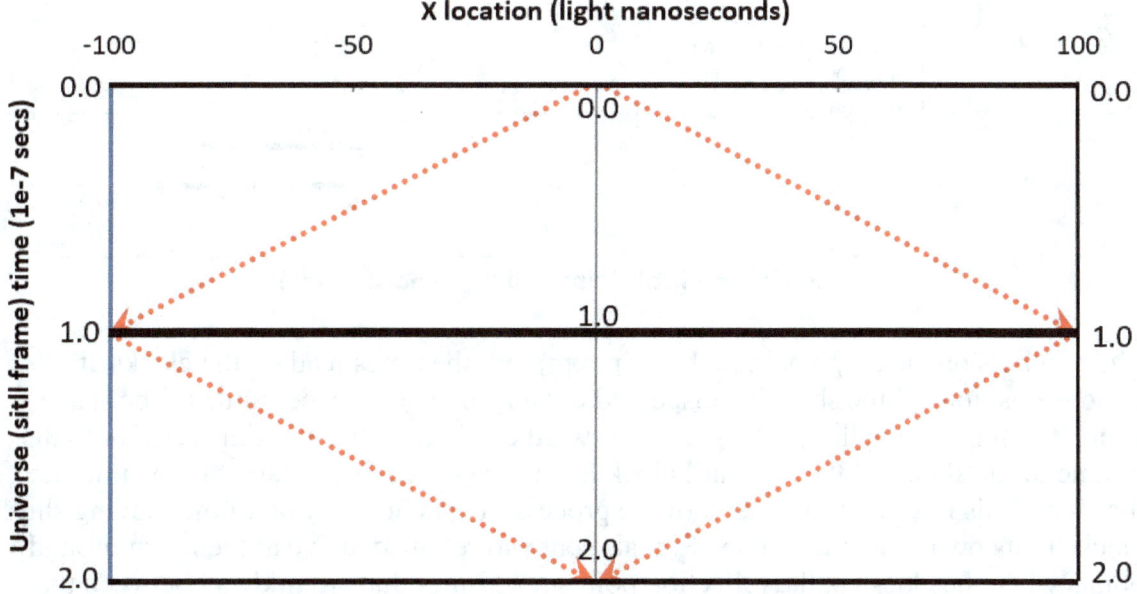

The above graph shows the trivial case of the ship that is not moving through space. It takes light 1e-7 seconds to travel from the ship center to the forward and the rear of the ship. The clocks on this ship are not only synchronized in the reference frame of the ship, but they are also synchronized in the "Now Moment" sense, that is, in every state of the universe, all the clocks on the non-moving ship always read the same time.

The graphic below shows settings of the clocks on a ship which is moving through space at the familiar very high speed of (v/c = .866, gamma = 2). This picture is consistent with the earlier modeling we did with 'bug swarms' for gamma = 2. Since the ship is moving rapidly to the right, it takes very little time (in State Universe) for light traveling rearward to get from the middle of the ship to the rear and equally little time to get from the front of the ship to the midship point. On the other hand, it takes light a relatively very long time (in State Universe) to get from the rear of the ship to the midship, or from the midship to the front of the ship.

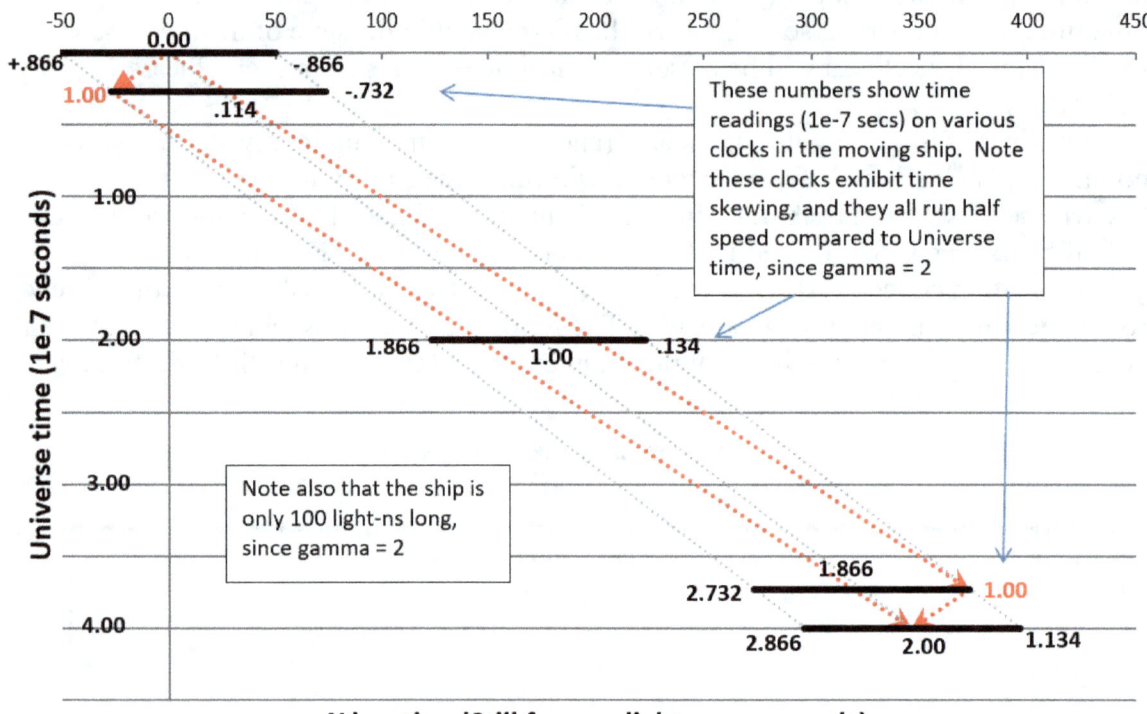

X location (Still frame - light nanoseconds)

The numbers on the graph next to the ship represent the times read on the clocks at various positions of the ship. As discussed earlier, in the reference frame of the space which the ship is travelling through, the forward clock is set back in time relative to the middle of the ship, and the rearward clock is set forward in time relative to the middle of the ship. This is an essential result of the process of synchronization of the moving ship's clocks in its own reference frame. Signals sent and returned arrive at their turn-around points when the clocks at the reflection points read times that are midway between the time the signal was sent and the time it returns to its source. The clock offsets are proportional to the distance of the reflection point from the source signal and are also proportional to the speed of the ship through space.

Note also in the graph that the "ticking rate" of the moving clocks is 1/gamma. In this case all clocks run half as fast as a clock that is not moving through space. So, for this gamma = 2 case, 4e-7 seconds pass in the Universe, but all the clocks on this ship advance only 2e-7 seconds. And note also that in the reference frame of space, the ship is half as long as the stationary ship, again, in concordance with our constructive model based on harmonious cycling of energy.

TAKING ON THE PARADOXES

Now I'm going to show a series of diagrams. The first picture will show the two ships in various positions, and the time on each clock on each ship will be shown. Then I'll re-draw the same picture with just certain times shown, so that we can more easily focus on how the symmetric measurements happen.

Each picture shows the two ships at 4 moments in State Universe, with time progressing forward from top to bottom within each picture. Note that the ship's lengths are not shown, but the moving ship is half the length of the stationary ship, because gamma =2. The only numbers shown are clock times at various positions in the ships. The stationary ship's clocks run at a 'normal' rate, that is, the fastest possible rate, since they are not moving through space, and they have no time-skewing. The moving ship's clocks run at half this rate, and they have time skewing.

The first moment shown is when the ships first start to pass each other. The blue ship is not moving, the brown ship is moving toward the right. At this moment, the forward clocks on both ships are set to read 0. Notice that for the blue ship, since it is at v=0, it has no time skewing, so its clocks all read the same time 'at the same time' in State Universe. The brown ship's clocks are skewed in State Universe, with rearward clocks set forward in time, so that in the brown ship's frame of reference, they are synchronized as shown in the previous section.

As the passing event occurs, all the clocks move forward in time, of course. At the second shown moment, the brown ship's nose is passing the middle of the blue ship. At the third moment, the brown ship's nose is passing the tail of the blue ship. And finally, at the fourth moment, the brown amidships clock is passing the tail of the blue ship. You will note that the time interval between the 4 positions is not equal; the time interval between the first and second position and the second third position is twice as large as the time interval between the third position and the fourth position. This is intentional, so that the presentations below can be accomplished with maximum efficiency.

If you question how all these specific times were arrived at from our constructive model, this will be explained in more detail in the next section where we demonstrate the Lorentz Transformation equations. In general, the advancement in time from step to step is a function of the relative velocity, and the time skewing is determined from the prior discussions regarding synchronization of clocks in objects moving through space.

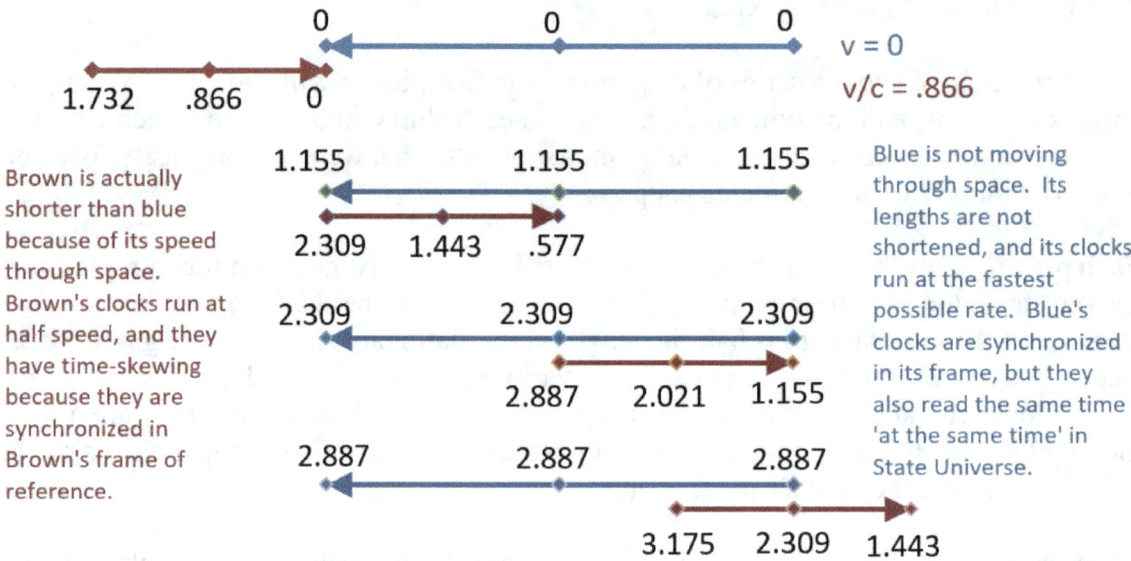

Clock times in Special Relativity, with one v=0 frame, Gamma = 2

You may want to study this first picture above; all the effects we have discussed in Special Relativity are can be seen in it. But to clarify one concept at a time, I've duplicated the above picture several times below, and in each case removed all the times except the ones which are crucial for understanding each symmetry phenomenon, one by one.

SYMMETRY OF MEASURED RELATIVE SPEED

Here's a symmetry that has generally been glossed over so far by me, and most other writers about relativity. With all the clock rate and length changes, and time skewing, it's not obvious at all that both observers would measure the speed of the other ship to be identically v/c in their own reference frames, but indeed it does turn out that way! In the example shown, both ships will measure that the other ship is moving at .866c relative to their own ship. Remember that it takes light takes 2 time units (1 time unit = 1e-7 seconds = 100 nanoseconds) to travel from one end of a ship to the other (200 light-nanoseconds distance). Both ships measure the time it takes for the nose of the other ship to travel their entire length as 2.309. So, the measured speed of the other ship is in both cases 2/2.309 = .866 as fast as light. Therefore, both ships measure v/c = .866 for the speed of the other ship relative to themselves. Again, the graph below is identical to the graph above, with only the relevant times shown for clarity of explanation.

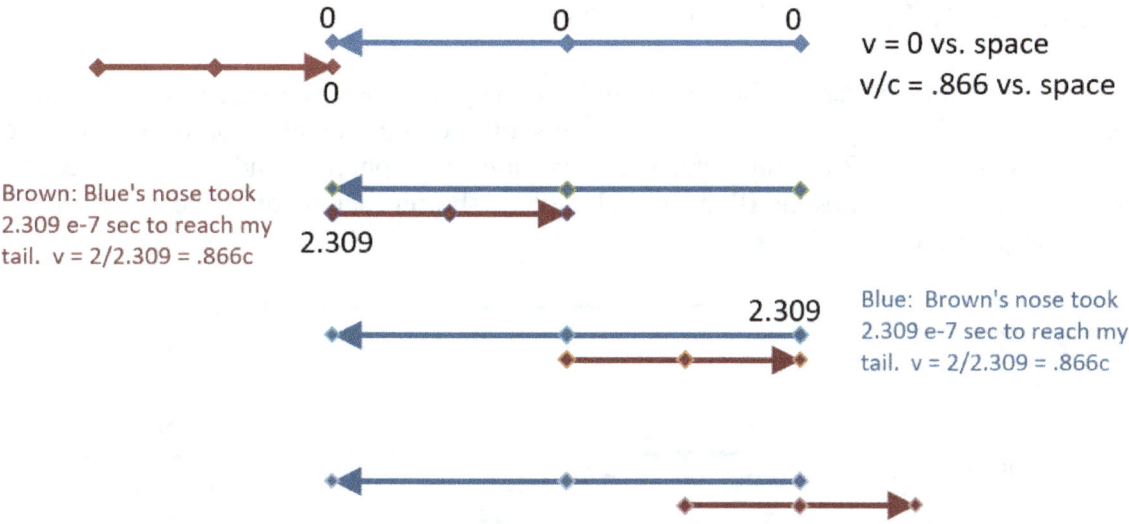

Symmetric Measurement of Relative Speed

SYMMETRY OF CLOCK RATES

We have already discussed this, but the following graphic shows how time skewing in the moving clock is the key to this so-called paradox. At v/c=.866, (gamma = 2,) both ships measure the other ship's clock as moving half as fast as their own.

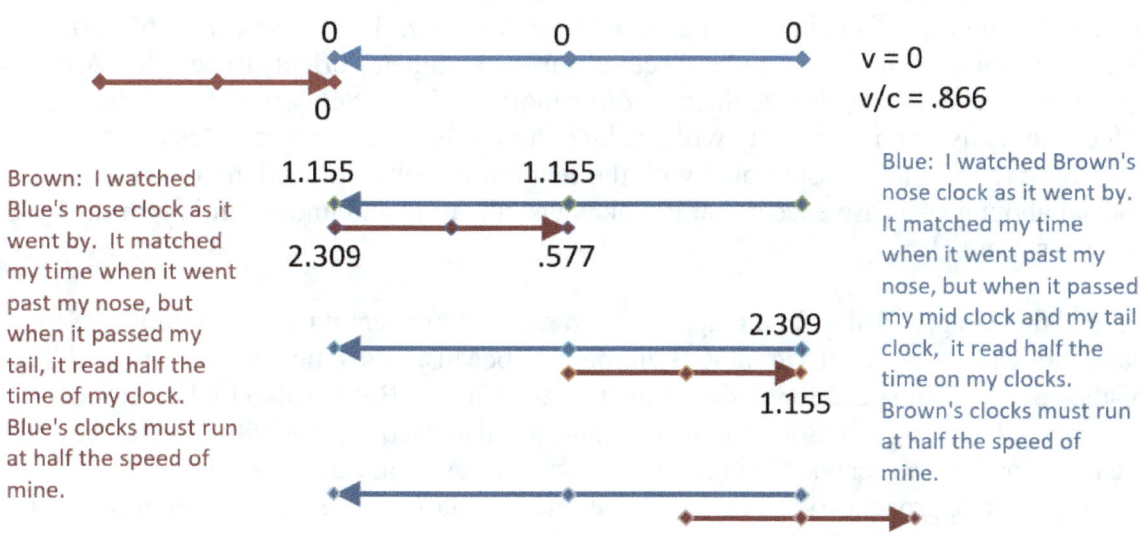

Symmetric Time Dilation

SYMMETRY OF LENGTH CONTRACTIONS

Although in State Universe, when moving through space at v/c=.866, gamma = 2, all of the Brown ship's energy patterns (masses) are shortenend in the direction of motion by a factor of two, and the Blue ship's masses at v=0 are fully spherical and not shortened, the Brown ship still measures the Blue ship's length as shorter than its own because of the Brown ship's time skewing.

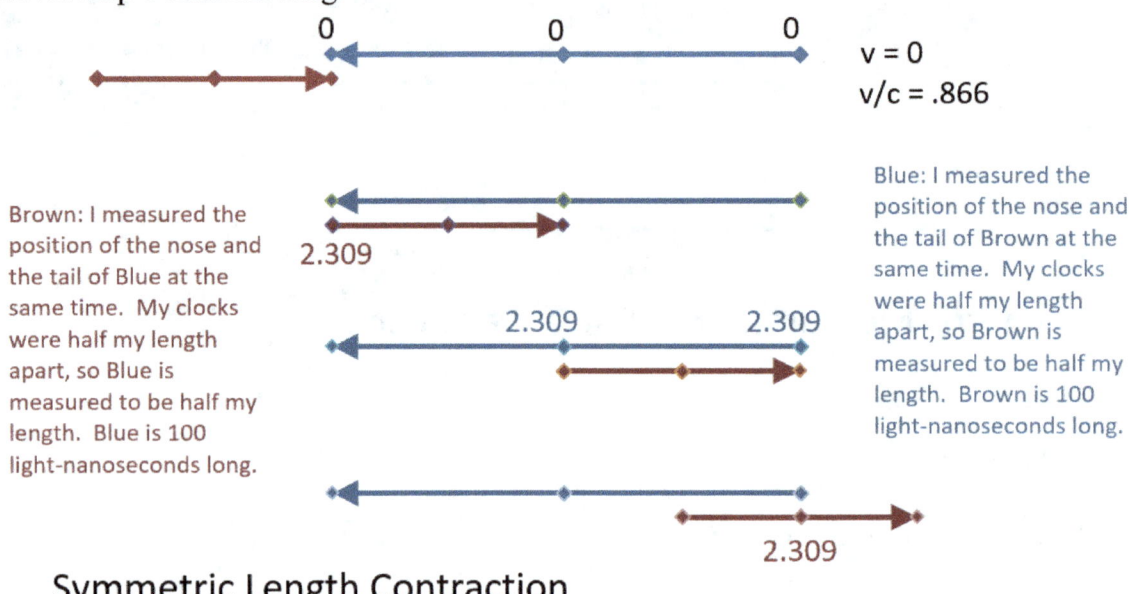

Symmetric Length Contraction

It is important to remember that my selection of v=0 and v=.866c relative to space was intentional but arbitrary. By this I mean that I chose these for the purpose of demonstrating the effects in the simplest and clearest way. But the selection of v=0 for one of the ships is absolutely NOT essential for the symmetry effects to be true. Any speeds for the ships which give them *relative* motion of v/c=.866 will produce these 2:1 effects, and any relative velocity will produce these effects to a larger or lesser extent, according to the gamma associated with the relative velocity. Indeed, there is nothing special about a v=0 case except that it makes visualization and understanding of the concepts more clear.

To end this chapter, I'll offer a graphic which shows how gamma varies with v/c. We have looked at the case of v/c=.866 (gamma = 2) because it's a nice convenient middle-of-the-road type of speed for studying relativistic effects. But we also looked at v/c = .5 (gamma = 1.15) in one bug swarm model, and we also used a v/c = .9897 which gave us a gamma of 7 in the second Twin Paradox example. As you can see from the curve below, as v/c approaches 1, gamma can become very large. The high powered physics accelerators get to speeds of v/c = .99999+, where gamma is a very large number indeed. Also clear is that for any speed that humans have personally attained, gamma is so close to 1.0 that relativistic effects are for practical purposes negligible. A trip to the moon slowed down the astronaut's clock by a small fraction of a second over the span of a week.

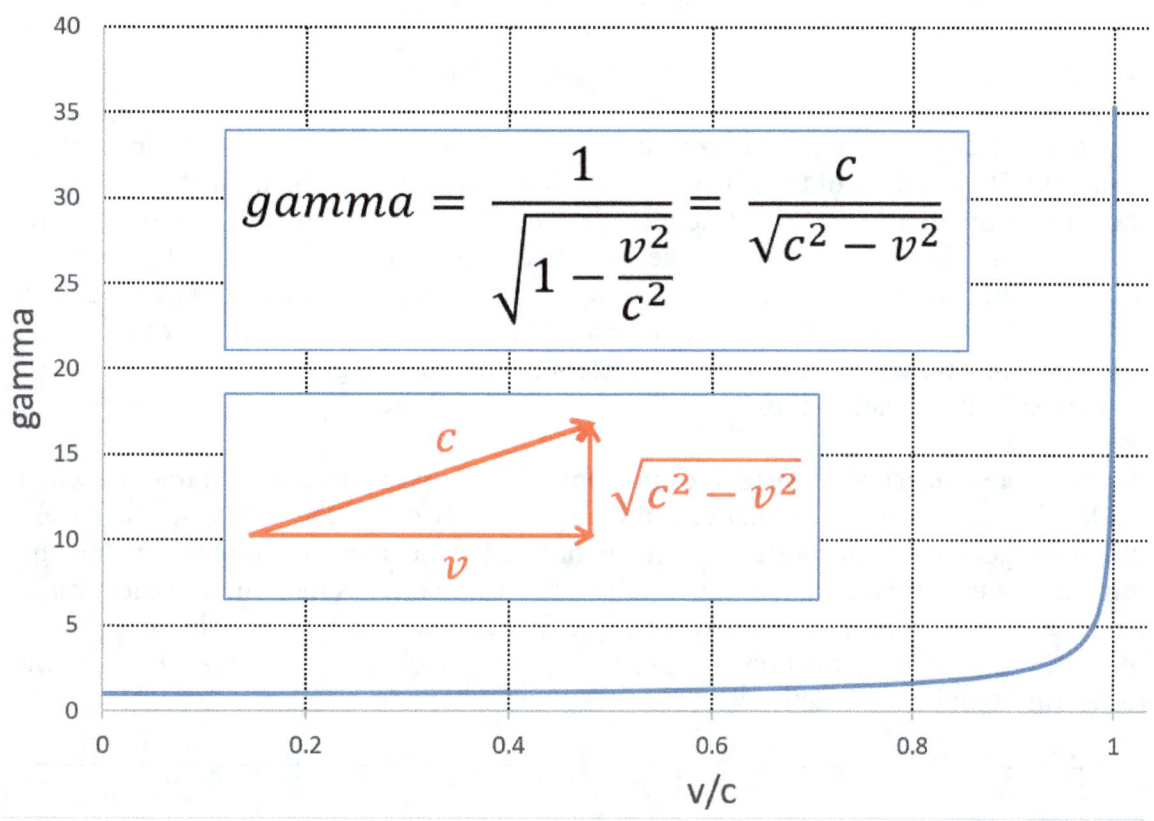

CHAPTER 12 - LORENTZ TRANSFORMATION EQUATIONS

Now that we have a constructive model in mind such that we understand what's physically happening, for reference, I'll present the Lorentz/Einstein transformation equations. Remember that Lorentz first derived these equations as relationships that must be true to fit the results of the Michelson-Morley experiments. Then, Einstein derived these same equations from his Principle of Relativity which was based upon the fact that the speed of light is measured to be the same in any constant-speed frame of reference. These equations are Einstein's first relativity result. Other ramifications regarding mass, including $E = mc^2$ were derived from these initial equations. The concept of spacetime, and Einstein's further work regarding accelerated motion and gravity were also built upon these initial relationships.

To completely match the Einstein formulation, I'll set up two reference frames exactly as he did. The direction of the relative motion is generally called the x direction, and each moving object has its own reference frame, one called the unprimed frame (x,y) and the other called the primed frame (x', y'). The relative velocity, as measured in each frame has an algebraic sign, which is positive (+v/c) if the 'other' frame is moving in the measuring frame's +x direction, and negative (-v/c) if the 'other' frame is moving in the measuring frame's -x direction.

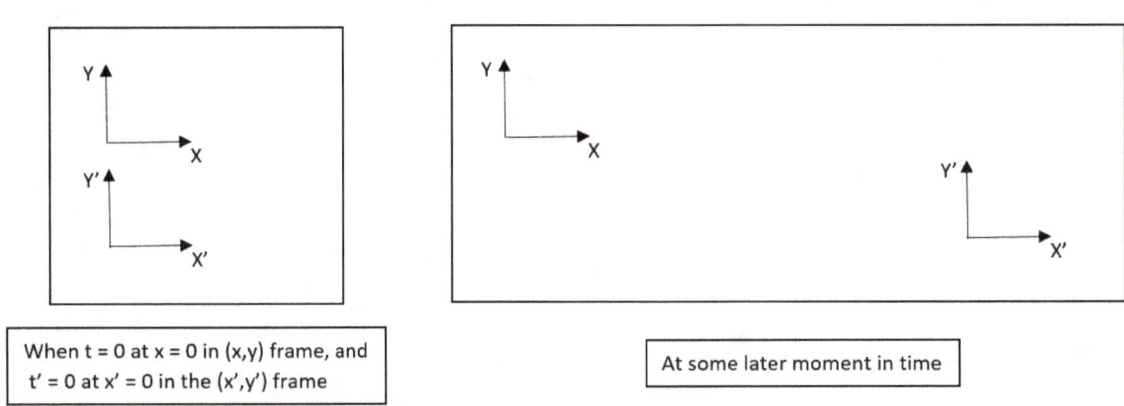

When t = 0 at x = 0 in (x,y) frame, and t' = 0 at x' = 0 in the (x',y') frame

At some later moment in time

For the two frames shown above, v, the relative velocity as measured by the unprimed frame is positive, because the primed frame is moving in the +x direction, but v', the relative velocity as measured in the primed frame is negative, because the unprimed frame is moving in the -x' direction.

Before looking at the Lorentz/Einstein equations, let's remember the Galilean/Newtonian transformation equations which were presumed to be true before Einstein's breakthrough:

$$x = (x' - v't') \qquad t = t'$$

$$x' = (x - vt) \qquad t' = t$$

These equations are intuitive and simple, but not perfectly correct. They become very incorrect as the speeds involved approach the speed of light, but they are extremely accurate when v is small compared to the speed of light. Remember that these equations show relationships between *measurements* made in each reference frame. For example, suppose an airplane (primed frame) passes a tower (unprimed frame) at a speed of 300 feet per second, and both tower and plane clocks read 0 when they pass. The relative velocities are measured as v = 300 ft/sec in the unprimed (tower) frame, and v' = -300 ft/sec in the primed (plane) frame. The unprimed (tower) frame, will *measure* the plane to be at x = +600 feet when the tower clock reads t = 2 seconds, and the plane's reference frame will *measure* the tower to be at -600 feet at t' = 2 on its clock. Formally, the calculations look like this:

At x' = 0   t' = 2   v' = -300
x = 0 − (-300)(2) = 600
t = 2

At x = 0   t = 2   v = 300
x' = 0 − (300)(2) = -600
t' = 2

Now here are the two fundamental equations of Special Relativity, originally set forth by Lorentz, which Einstein later re-derived from his formal Principle of Relativity in his 1905 paper:

$$x' = \gamma(x - vt) \qquad t' = \gamma\left(t - \frac{vx}{c^2}\right)$$

$$x = \gamma(x' - v't') \qquad t = \gamma\left(t' - \frac{v'x'}{c^2}\right)$$

Where $gamma = \gamma = \dfrac{1}{\sqrt{1-\frac{v^2}{c^2}}}$

v = relative speed in the x direction as measured in the unprimed frame
v' = relative speed in the x' direction as measured in the primed frame
x and t are location and time measured in the unprimed frame
x' and t' are location and time measured in the primed frame
c is the speed of light (always meaured to be the same in either frame)

Notice that when (v/c) is small, the Lorentz equations reduce to the Newton equations because v/c is small, and gamma is nearly one. For example, an Apollo ship's maximum speed, the fastest humans have ever traveled relative to Earth, was 11 kilometers per second. This speed has v/c = .0000367, and gamma = 1.00000000067   For relativistic effects to be significant, much, much higher speeds are needed.

I'd like to show how the Lorentz transformations work and show that these transformations are consistent with the constructive model we've developed. For this, I'll present two cases. First, we'll look at the case we are most familiar with, namely, when one frame is at v = 0, (not moving through space,) and the other frame is moving through space at .866c. The relative motion of these two frames is, of course, .866c.
For the second case, we'll also let the relative velocity be .866c, but we'll have both frames moving through space - both frames will experience length contraction, time dilation, and time skewing. We can build both these cases using our constructive model, then confirm that the Lorentz transformations agree with these models.

The graphics we'll see here will all be shown from a State Universe point of view, in other words, with our constructive model in our mind's eye, we will see what is going on physically, not just mathematically. Remember that anyone in either frame would be unable to tell if they were moving through space or not. Although that is true, we are free to consider a v = 0 case, or any other physical scenario we wish.

CASE 1 – one frame stationary in space, other frame moving through space

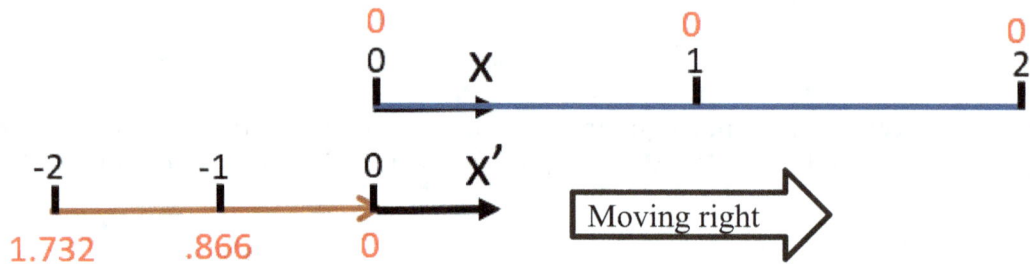

This picture should be familiar from the earlier chapter which discussed paradoxes. In that example, we had identically constructed ships that are 200 light-nanoseconds long. The lengths of the ships were shown as 200 light-nanoseconds = 2e-7 light-seconds = 2 length unit in these pictures. Clock times, shown in red here, are similarly in units of 100 nanoseconds = 1e-7 seconds = 1 unit of time in our pictures. However, these units could as well be time in years and distance in light-years, or time in minutes and length in light-minutes, and so on. When we use length units like this, we are 'normalizing' c. By this I mean that in these units, c = 1. Light travels one unit of length per one unit of time. This greatly simplifies the numerical analysis. If distances in meters are ultimately desired, the light-year, light-minute or light-(1e-7)-second can be converted to meters.

Per our constructive model, the non-moving blue ship has all clocks reading the same time in State Universe, as well as being synchronized in its (unprimed) reference frame. The brown ship, which is moving to the right at .866c through space, has time skewing equal to v/c time units per unit length unit in the (primed) reference frame of the moving ship. Since the reference clock at x' = 0 is set to t' = 0 as the ships first pass, its other (trailing) clocks are set forward in time in the State Universe sense.

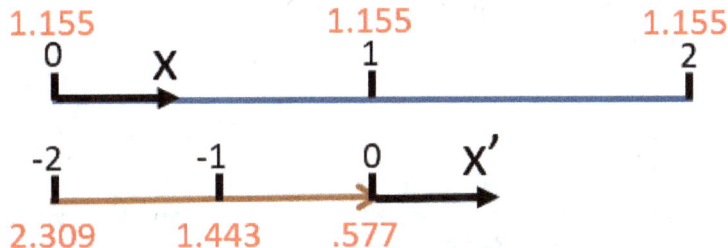

Now here is the same system a short time later. Let's think for a moment about how the times here are obtained from our constructive model. Notice first that in this State Universe view, the moving ship is half as long (gamma = 2) as the non-moving ship. From the point of view of the unprimed v = 0 frame, the nose of the other ship has traveled a distance of one length unit in its frame (from x = 0 to x = 1) at a relative speed of .866c. Since light traveling at speed c takes 1 time unit to travel 1 length unit, and the nose of the brown (unprimed) ship is only moving at .866c, it takes the nose of the ship 1/.866 = 1.155 units of time to travel the same 1 length unit. Therefore, the time t at x = 1 is 1.155, and all the other clocks in the non-moving frame are the same because there is no time skewing in a non-moving frame.

Similarly, at this moment, the brown (primed) ship sees that the nose (left end) of the blue ship has traveled the full length of its own ship, in other words, has traveled 2 length units in its frame of reference. Since in the primed frame light takes 2 time units to travel the same 2 length units, and the relative speed is .866c, the left end of the blue ship takes 2/.866 = 2.309 time units to travel the same distance. Therefore, at x' = -2, t' = 2.309. Note that from our State Universe point of view, the clocks on the brown, (primed) moving ship are advancing at half the rate of the clocks on the non-moving ship, but the time skewing remains the same, .866 time units per length unit in its frame. This is how we arrive at the other clock times in the primed frame. (2.309 - .866 = 1.443 and 1.433 - .866 = .577)

Now, let's look at the Lorentz Transformations and show a couple examples for this case. These equations find (x', t') in one frame that correspond to (x, t) in the other frame.

$$x' = \gamma(x - vt) \qquad t' = \gamma\left(t - \frac{vx}{c^2}\right)$$

For the blue, unprimed (x, t) frame the relative velocity v of the other frame is +.866c, because the other frame is moving in the +x direction. Remember that in the units we are using, the speed of light has been normalized, so c = 1.

At: x = 1   t = 1.155   v = .866c = .866   gamma = γ = 2
x' = 2 *(1 - (.866)*(1.155)) = 2 * (1 – 1) = 0
t' = 2 * (1.155 - .866 * 1) = 2 * (.289) = .577
(x, t) = (1, 1.155) corresponds to (x', t') = (0, .577), which agrees with our graphic.

The Lorentz transformations work in both directions, (from unprimed to primed, and from primed to unprimed) but remember that from the point of view of the primed frame, the 'other' frame is moving in the -x' direction, so v' has a negative sign. The equations look like this:

$$x = \gamma(x' - v't') \qquad t = \gamma\left(t' - \frac{v'x'}{c^2}\right)$$

At: x' = -1   t' = 1.443   v' = -.866c = -.866   gamma = γ = 2
x = 2 *(-1 – (-.866)*(1.443)) = 2 * (.25) = .50
t = 2 * (1.443 – (-.866)*(-1)) = 2 * (.577) = 1.155
(x', t') = (-1,1.443) corresponds to (x, t) = (.50, 1.155), which agrees with our graphic.

CASE 2 – both frames moving through space

Now, let's look at the second case, in which both frames are moving through space, and neither frame is in a v = 0 state. Although our constructive model and the Lorentz Transformations work for all relative velocities, let's take another case where v/c = .866 and gamma = 2. Given that v/c = .866, there are an infinite number of scenarios for "where space is", but let's suppose that both frames are moving at the same speed through space. What speed through space should they both have so that their velocity relative to each other is .866c?

We'll again use the '3 frame' equation of relativity which we used earlier in the book when setting up the second Twin Paradox case.

Special Relativity gives us: $\quad S = \dfrac{(v+u)}{\left(1+\frac{v*u}{c^2}\right)}$

S is the mutual relative speed of two frames when one frame moves relative to space at speed v, and the other frame moves relative to space with speed u. If we set u and v to be equal, and we want S to be .866 ($=\frac{\sqrt{3}}{2}$) we have:

$$\frac{\sqrt{3}}{2} = \frac{2u}{(1+u^2)} \quad \text{or} \quad u^2 - \left(\frac{4}{\sqrt{3}}\right)u + 1 = 0$$

which we can solve with the quadratic equation: (we use the root that is less than 1!)

$$u = \frac{\frac{4}{\sqrt{3}} +/- \sqrt{\frac{16}{3} - 4}}{2} = \frac{2}{\sqrt{3}} - \frac{1}{\sqrt{3}} = \frac{1}{\sqrt{3}} = .57735$$

For this speed relative to space, the gamma *relative to space* is $\sqrt{1.5} = 1.225$

So, let's use our constructive model to set up two reference frames, one moving left at .57735c and one moving right at .57735c. From a State Universe perspective, both these frames will be shortened by the ratio $\frac{1}{\sqrt{1.5}} = .8165$, and their time skewing will be .577 units of time per length unit in their frame. In both ships, the time skewing is such that the trailing clocks are set forward in time in the Statue Universe sense. As in the prior case, both reference frames have their +x axes pointing to the right. The unprimed frame is moving left, and the primed frame is moving right. Here is the situation when the noses of the ships first pass:

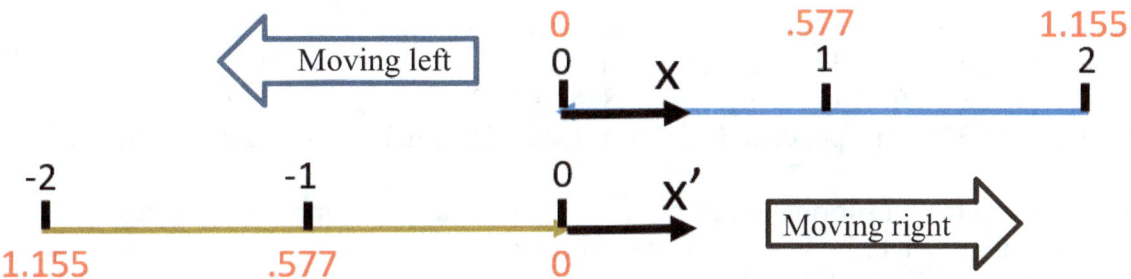

Next, let's use our constructive model and the same logic we used in case 1 to show the times on all the clocks at a later time. We'll let the frames move to a side-by side position. Remember that relative to space, the gamma for these frames is 1.225, but relative to each other, v/c = .866, and the gamma is 2.

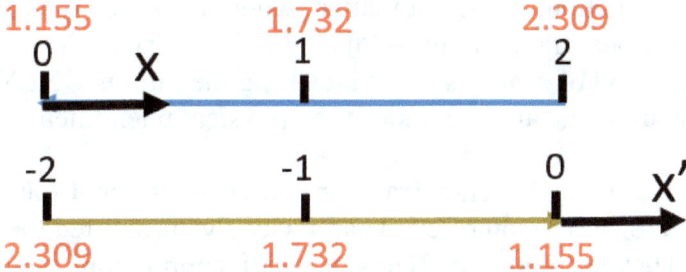

Each ship passed nose to nose at t = 0 and t' = 0 on their clocks. Because their mutual relative speed is .866c, when the nose of the other ship reaches their own tail, that nose has traveled 2 length units in their own frame in 2/.866 = 2.309 time units. In other words, light would travel from their own nose to their own tail in 2 time units, but the nose of the other ship is going slower than light, so it takes longer than 2 time units for the other's nose to traverse their own length. The clock time at the tail of both ships is therefore 2.309. With the time skewing maintained at .577 per length unit, the times on the other clocks are computed. Since the other clocks are *leading* the tail clocks through space, they are set back in time (in State Universe) relative to the tail clocks. Remember, however, that in the reference frame of each ship, all the clocks are synchronized in the Einstein sense. No one on either ship can detect their speed through space, so they can't know that their clocks might be running slower nor their lengths shorter than they would if they were not moving through space. Nor can they know if or how much their clocks might be 'reading different times' at the same time in State Universe.

Now, let's again plug into the Lorentz equations to confirm their agreement with our constructive model for this case.

At: x = 0   t = 1.155   v = .866   gamma = γ = 2
x' = 2 *(0 - (.866)*(1.155)) = 2 * (0 – 1) = -2
t' = 2 * (1.155 - .866 * 0) = 2 * (1.155) = 2.309
(x, t) = (0,1.155) corresponds to (x', t') = (-2, 2.309), which agrees with our graphic.

At: x = 1   t = 1.732   v = .866   gamma = γ = 2
x' = 2 *(1 - (.866)*(1.732)) = 2 * (1 – 1.5) = -1
t' = 2 * (1.732 - .866 * 1) = 2 * (1.732 - .866) = 1.732
(x, t) = (1,1.732) corresponds to (x', t') = (-1, 1.732), which agrees with our graphic.

And, running the transformation from the primed frame to the unprimed frame:
At: x' = 0   t' = 1.115   v' = -.866   gamma = γ = 2
x = 2 *(0 – (-.866)*(1.115)) = 2 * (1) = 2
t = 2 * (1.115 – (-.866)*(0)) = 2 * (1.115) = 2.309
(x', t') = (0,1.115) corresponds to (x, t) = (2, 2.309), which agrees with our graphic.

Readers can try more cases if they wish, but what they will find is that it doesn't matter "where space is" – for all cases in which the relative velocities of the two frames is .866c, the Lorentz Transformations are correct. And that was Einstein's point. For the purposes of calculation and prediction, it doesn't matter "where the space is." That's why the concept of the "wave bearing medium" – "the ether" – or "space" was eliminated from relativity. Consideration of space as a wave-bearing medium is ONLY valuable if we want to visualize and understand the underlying physical phenomena.

When I first encountered the Lorentz Transformation equations, I found them a bit difficult to get used to, and to interpret. That's why I've presented them here near the end of this book. They show all the effects of length contraction, time dilation, and time skewing that we've been discussing, of course. Stepping through Einstein's mathematical derivation of these relationships (from the premise that the speed of light will be measured to be the same regardless of the motion of one's reference frame) is not very difficult – it only took him two pages to do it. It was an intellectual master stroke, but the algebra involved is not that difficult.

However, as an engineer, I was used to following derivations from premises that made physical sense in the first place. When you do that, the various parts of the derived equation are often easy to interpret. Somehow, I forgot that in the derivation of the equations of Special Relativity that the premise contained the mystery. I bought the premise, but I didn't really understand it. So, it was difficult to understand what each term in these derived Lorentz Transformation equations represented, physically. Length contraction and time dilation are obviously embodied in gamma, but at first, I had little understanding of the physical meaning of the equation for gamma. The time skewing is represented in the $\frac{vx}{c^2}$ term, but without a constructive model, I couldn't recognize it as such. Now we know that time skewing is clearly proportional to $\frac{v}{c}$ and is also proportional to x, the position in the skewed frame. Note that the time skewing term has a negative sign; this represents the fact that in each frame, the forward (+x) clocks are set *back* in time, and trailing (-x) clocks are set forward in time in the 'State Universe' sense as we have been discussing it. This is what I mean by being able to interpret the various parts of the derived equation.

The reader should note that this book has not covered all the topics Einstein developed in his works on Special Relativity. Notably, the relativistic effects related to mass and momentum have not been touched upon here. But these follow from Einstein's original derivation of the Lorentz Transformation equations from the Principle of Relativity. Hopefully the addition of this constructive model as a "prequel" to Einstein's derivation serves to help readers feel comfortable with the physical meaning of the phenomena of relativity.

AFTERTHOUGHT - PULSARS: AN ABSOLUTE TIME REFERENCE?

The idea that an absolute reference cannot be found is so widely accepted that the proposal of an experiment to measure one's absolute speed through space is an invitation to absolute ridicule. To propose such an experiment would probably get you laughed out of the University, kicked out of Grad School, or fired from your research job. Certainly your 'Book on Relativity' would be laughed off the shelves. But wait. Try as I might, I cannot find the flaw in a simple thought experiment which would determine one's absolute speed through space by locating a local v=0 frame. Fortunately, this is just a thought experiment, not a real, practical experimental proposal. That makes me feel safer.

Pulsar pulse rates are extremely steady. The most accurate clocks on Earth are now timed by reference to microsecond pulsars. The steady pulsar rates are being used to overcome problems that atomic clock have with non-constant solar distances, wobbling Earth orbits, planetary influences, and other miniscule effects. These subtle things change the velocities and gravitational fields of the atomic clocks, which alter their running speed slightly, leading to small inaccuracies. Remote pulsars are not affected by such local events; they emit very steady pulses over long periods of time. Many of these pulsars emit blips of energy thousands of times per second – these are the microsecond pulsars.

To "find" v=0 space, first, you would want to get to a remote sector of space far from significant gravity. Then the goal would be finding three microsecond pulsars with stable frequencies in orthogonal positions which would provide the opportunity to maximize the clock rate of the atomic clock you have onboard your ship by changing your ship's velocities in the three coordinate directions so as to minimize all three measured pulsar rates. The measured frequency of the pulsars would be slowest when the absolute rate of your own clock is maximized. This would put you in a v=0 condition, i.e. 'still in space.'

This is of course, not a practical exercise by any means. The only purpose of this thought experiment is to challenge the notion that there could be *no way* to detect one's absolute motion through local space. Since there is no escaping the fact that one is IN the Universe, there is no escaping gravity. That could only happen in a universe without mass, and that would not be a very interesting universe, particularly since you and your clock and your ship would not exist.

THOUGHTS ON THE SOCIAL IMPLICATIONS OF RELATIVITY

"Through multiplication upon multiplication of facts, information, theories and hypotheses, it is science itself that is leading mankind from single absolute truths to multiple, indeterminate, relative ones. The major producer of social chaos, the indeterminacy of thoughts and values that rational knowledge is supposed to eliminate, is none other than science itself."

                        Robert Pirsig, <u>Zen and the Art of Motorcycle Maintenance</u>

Things are going to slide… slide in all directions
Won't be nothing,
Nothing you can measure anymore.
The blizzard, the blizzard of the world
has crossed the threshold
and it's overturned
the order of the soul.
When they said REPENT, REPENT!
I wonder what they meant…

    - Leonard Cohen, <u>The Future</u>

"It depends on what the meaning of the word 'is' is."
    - Bill Clinton, 1998

"Don't be so overly dramatic about it, Chuck,
you're saying it's a falsehood, and our press secretary,
Sean Spicer, gave alternative facts to that."
    - Kellyanne Conway, January 2017

"Truth is relative…
He may have a different version of the truth than we do."
    - Rudy Giuliani, May 2018

When the news stories get me down
I take a drink of
Freedom to think of…
My America from toe to crown.
It's never long before
I know just why I belong here.

    - Buffy Sainte-Marie, <u>Soldier Blue</u>

The new reality according to 20th Century Science has been that there is no absolute reality, only definable relationships between different realities. To Science, a multiplicity of valid measurements implies a multiplicity of truths. One concern I have is the idea that this philosophical concept has spread into the culture at large.

When Physics draws new fundamental conclusions about reality, there is a time lag before its new ideas permeate the culture. Newton first published his Principia in Latin in 1687, later in English in 1728. Freudian psychology was developed at the end of the 19th century, just before the development of Special Relativity in 1905. The language of Freudian psychology is loaded with Newtonian concepts from the 18th Century. According to Freud, people had impulses, and drives. Psychological conflicts were viewed in terms of a balance of countervailing forces. People took actions and had reactions to things. Patients were said to be resistant to change. Life events had emotional impacts. These are all Newtonian concepts.

A similar Newtonian mindset continued to dominate business, economics, and political spheres throughout the first half of the 20th century. It became somewhat more sophisticated, but it was still fundamentally Newtonian. For example, Newtonian principles of equilibrium became more sophisticated when applied to complex systems such as the economy. Complex mechanical, chemical, economic and social systems may be stable, or may move toward states that are increasingly unstable. Rational analysis can predict, or at least explain, rapid, catastrophic jumps between two very different operational states of a normally stable system.

Now that relativity has been in place for over a century, what do we have? What language is used when we talk about psychology, history, and culture?

In the 1950's, the culture was still generally operating on a foundation of assumed absolute truths. The Newtonian concepts were still strong. People were given electroshock and had their personality erased if they didn't conform and accept the common view of reality. But by the late 1960's this had changed. There were different strokes for different folks, and encouragement to "do your own thing" was the order of the day. In the 1970's, we had pop psychology proclaiming "I'm OK, you're OK." Culturally, reality became more fluid. It was said that "you create your own reality." Was this the beginning of cultural relativity?

In the mid 1980's I had a shocking and very revealing argument with a well-respected, board-certified psychologist who claimed with complete seriousness that when people disagreed about basic facts concerning 'what happened,' in a specific situation, it did not imply that anyone was wrong. I argued that there was only one set of correct facts, and that the different viewpoints expressed implied mistaken observations, incorrect memories, or outright lies. The psychologist claimed that despite the contradictions, everyone could be perceiving correctly, and telling the truth, and everyone could be 'right' - they simply existed in different realities; everyone had their own personal truth.

This was shocking at the time, but now I get it. Relative truth is now the cultural norm. Relativity worked its way into the culture. It seems that everyone now has their own story, their own perspective, and their own unique history, and there are very few basic facts on which everyone can agree. Of course, this can lead to problems when a large number of people live in a fantasy world of their own design. You can't stop the dreaming, but it's hard to explain to a kid who rejects logic that they *probably* won't be famous someday as a sports star or popular performer. When did "he thinks the world revolves around him" start to lose its power as an insult? When did narcissism become a feature, not a bug?

One obvious trend in modern American culture is an increasing level of political disagreement over basic facts. Rather than sharing a set of commonly agreed-upon facts and discussing what approach should be taken to solve problems, we see an increasing level of disagreement over what the basic facts are. We have political fact-checkers, and people who check the fact-checkers. And we have multiple fact-streams in the form of politically biased network "news" programming and websites. The battle for minds and souls has moved from the realm of interpretation of facts to the realm of definition of facts. I see an erosion of the highest intellectual value: truth.

Is truth really relative? Or is there an absolute truth? Do I think that opening a small crack into to consideration of an absolute reality in relativity theory will actually change anything socially? No, I don't expect that. I want to emphasize, however, that I am not arguing for a return to pre-Einsteinian, Newtonian absolutism, a clockwork universe in which everyone should agree. People DO have their own unique points of view, and that makes life a potentially rich experience. I am suggesting instead that a concept of indeterminate absolutism, where no one's point of view represents the truth, is superior to the relativistic concept that multiple views all represent the truth with equal validity. We might all get along better if everyone stopped insisting that they were sure they were right.

We should all be able to agree on two fundamental ideas:
1. There is one absolute truth.
2. None of us can know it.

The distinction between "different perceptions of reality" and "different realities" may seem linguistically subtle, but it is intellectually crucial. What change in the cultural foundation might a new perspective bring? Individual experiences of reality would still be viewed as various and unique. But I would hope for an increased awareness and acceptance that no one can claim a correct lock on truth. We are each what we are because of what we've been through. There is an absolute reality, and we are all a part of it, we all take part in it. But each of us sees it from our unique perspective. If we can share our different experiences, and listen carefully to each other, we can search for a better version of the truth – together. We might have more respect for those who disagree with us, knowing that our view is simply our view. We might also be more humble, and cautious in situations where we share a strong sense of agreement with others, realizing that the absolute truth in any situation is unknowable.

Physics has many aspects, and it's well known that one line of thinking in Quantum Mechanics says that if a thing can happen, it does happen, which leads to a mind-boggling multiple universe theory in which all quantum probabilities possible are played out – somewhere in an infinity of split-off universes. I'm not the only one who has philosophical issues with this. But even if it is true that the Universe does run multiple shows, surely at any given moment we are only involved in one of them. The fact that no one can rightfully claim to know the absolute truth is a part of the nature of reality itself.

www.ingramcontent.com/pod-product-compliance
Lightning Source LLC
Chambersburg PA
CBHW060424220526
45465CB00008B/3005